Homebuilt Airplanes

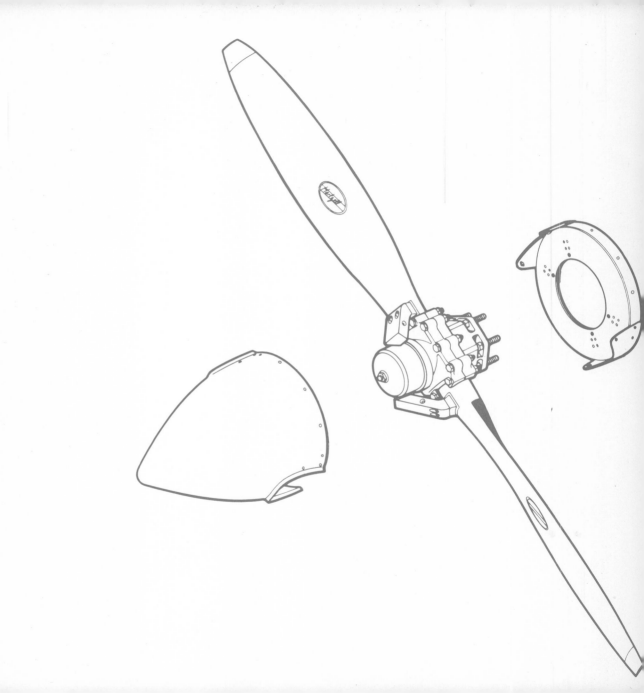

Homebuilt Airplanes

Photography by
Baron Wolman

Text by
Peter Garrison

Design by
Clyde Winters

A Prism Edition
Chronicle Books
San Francisco

The majority of the photographs in Homebuilt Airplanes
was exposed on Kodachrome-64 and Tri-X film in a Nikon-
FE automatic camera with MD-11 motor drive. Air-to-air
photos were made from the right seat of a Cessna-172
chase plane. Additional photography by Peter Garrison,
Paul Garrison, George Larson and Stephan Wilkinson.
Special thanks to Frank Christensen of Christen Industries
for the line drawings, and to Jack Cox of the Experimental
Aircraft Association for the historical photographs.

Produced and directed by Baron Wolman/SQUAREBOOKS,
Post Office Box 144, Mill Valley, CA 94941.

Edited by Elaine Ratner and Tim Ware.

Production supervision by Jennings Ku.

Color separations by Color Tech, Redwood City, CA.

Typography: Design & Type, Inc., San Francisco, CA.

Printed by Kingsport Press, Kingsport, TN.

Printed in the United States of America.

Library of Congress Cataloging in Publication Data

Garrison, Peter
    Homebuilt airplanes.

    "A Prism Edition"
    1. Airplanes, homebuilt. I. Title.
TL671.2.G29   629.133'343   79-17448
ISBN 0-87701-149-4

Chronicle Books/Prism Editions
870 Market Street
San Francisco, CA 94102

After years of exposure to the world of homebuilt airplanes, we knew it deserved a book. We knew, too, that we could hardly do justice to the astonishing variety of homebuilt designs and designers, the luxuriance of types, the extravagances of craftsmanship and detail, or, for that matter, to the technical achievements—and failures—of amateur builders. Reluctantly, we had to concentrate our attention upon a few, and let them speak for the many.

A small selection of samples from among a large number of possibilities is bound to meet objections. Surely one airplane was omitted that should have been included, or another was included that could have been omitted. Perhaps there are too many of one sort, not enough of another. The author may be biased. The photographer might have missed the best angle. And so on.

Knowing that we couldn't please everyone, but hoping at least to please many, we tried to select from the vast choice of homebuilt airplanes ones that were typical of different points on the spectrum, ones that had some larger significance, either as representatives of a genre or as milestones in the history of the homebuilding movement. In some cases the choice of a particular airplane was even influenced by time, geography, and the possibility of getting the airplane, the photographer and a photo plane together at one time in suitable weather.

With apologies, then, for its necessarily limited scope, we offer this little book as a gateway to a much larger world.

1 | INTRODUCTION

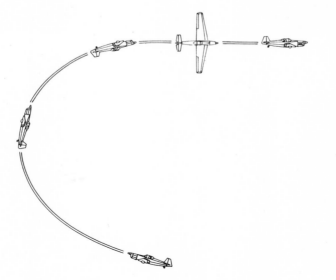

Of the pilots who have come to the airport this Saturday morning, there is one for whom the visit has a special significance. He escorts his guest with an air of stagey nonchalance; fumbles unusually long with the lock on his hangar door; and as he pushes the door open keeps his eyes on the ground, seemingly intent on his effort, but really not wishing to distract the visitor's attention from the spectacle he is unveiling. It is only an airplane—but an airplane unusually small, brilliantly painted, beautifully polished. As he stands by it the pilot pretends to be distracted by some flaw on the shining surface of the wing, which he carefully buffs out with a cloth—though, to the guest's eye, there is no flaw there. The airplane is uncannily clean, and the hangar around it orderly, everything put away so that no stray object intrudes upon the space which the little airplane fills with its curious glow. One could almost suspect that the pilot had come out here the night before just to put things into this perfect order, to set the stage for this morning's presentation. The airplane's glow is really the glow of reflected pride; because this is, in fact, not an ordinary private plane. It is a homebuilt—an airplane which the pilot built with his own hands.

To fly an airplane is to enter into a magical terrain, unexplored until a few decades ago. How maddening it used to be that birds and even despicable insects could do this thing that men could not. As people thirsted vainly for the

power to fly, they conferred it upon their gods, their angels, their magicians and their heroes.

Now flying is no longer rare. The techniques, it turns out, are fairly simple. For each of us who flies there is that first exhilarating experience when we fully and suddenly grasp the wonderful beauty of it, the fear and triumph with which humans scale the sky. After a while that fine sensation is nibbled away by habit, until we bury ourselves in mystery novels rather than look out the window of an airliner.

People who build their own airplanes are trying to preserve the childish wonderment of their first discovery of flight. To fly and to have made the flying machine oneself puts one into a rather select company. Like the fabulous artificer Dedalus—who should be the patron saint of people who build their own planes—the builder has launched himself into heaven; in a metaphorical sense, he has transcended human limitations. Airplanes stand for power and freedom; to have made one is to have made oneself powerful and free.

This is not to say that people who build airplanes are super-human, or for that matter that they are unusually happy. But, like a war hero or a champion athlete, they have the special satisfaction of having executed a splendid act. They have created for themselves a small, generally unrecognized, but ineradicable championship.

The U.S.A. is the world capital of this curious hobby, followed by a few other countries—France, Australia, South Africa. Certain conditions must exist to make homebuilding airplanes possible, principal among them general affluence, technological advancement and, rarest and most arbitrary, a liberal regulatory climate. In the United States, where all these requirements are fulfilled, the "homebuilt movement," as its members call it, has grown most rapidly. Homebuilts—homemade planes—in operation in the U.S. now number more than 6,000; thousands more are being built and the selling of airplane plans and kits has become a lucrative business. In response to the growing popularity of the pastime (the Experimental Aircraft Association—or the EAA as it is commonly known—has over 60,000 members) new designs have been developed that can be built rapidly, inexpensively, and without excessive skill or knowledge. Partly because of the commercial impetus this market has provided, and partly because the regulations governing the creation and use of homebuilts permit tremendous latitude for experimenters, the homebuilt movement has become the principal source of innovative design in the field of private aviation. The flow of ideas between industry and hobbyist has reversed direction in the last few years, and now the homebuilders have many things to teach the manufacturers.

In the early days of aviation all airplanes were amateur-built by experimenters whose names are now famous: Lilienthal, Wright, Santos-Dumont, Bleriot, Curtiss. But the modern homebuilder is different. Although his airplane is classified by the Federal Aviation Administration as "Experimental," the name is often inaccurate. Only a few homebuilders are breaking new ground. Most are building copies of proven designs, incorporating a few personal modifications, and putting on their airplanes the stamp of their own workmanship and style. They are, for the most part, not experimenters but craftsmen.

The genealogy of the modern homebuilt begins with the small personal biplanes of the thirties and the Goodyear Racers in the National Air Races of 1948 and thereafter. In the renaissance of racing in the sixties, the Goodyear Racers were supplanted by Formula Ones, in an attempt to create a class for which many people would be able to build competitive airplanes. One of the most popular and successful of the early homebuilt designs was the Wittman Tailwind, designed by Steve Wittman, whose homely racers were fashioned with an unschooled but unerring intuition. Other favorites were the Long Racer, which came to be known as the Midget Mustang after the beautiful P-51 Mustang fighter (to which it bore almost no resemblance), the Smith Miniplane, and the Knight Twister.

Factory-built airplanes in the sixties were without exception dull, uniform, functional and impersonal to an even greater degree than cars. And early homebuilt designs achieved goals which factory planes could not:

They were accessible.
Factory planes have always been expensive, but their price had risen out of proportion to the cost of living. The postwar fantasy of an airplane in every driveway folded before 1950; but its impetus survived in the then tiny homebuilt movement. The earliest designs were simple, small and as cheap to build and propel as was possible.

They had special, but limited capabilities.
Homebuilts originally followed a pattern of rusticity. Many had open cockpits. All had fixed tailwheel-type landing gears and fixed-pitch props, and were capable of being flown from rough grass fields and put through aerobatic maneuvers. In this respect they reflected a mentality about twenty years behind the times, but one considered a virtue, because modern airplanes, besides being too costly, were seen as cumbersome and soulless.

They were small.
Small not only because smaller airplanes take less time and money to build, but also because more performance could be obtained from a small airplane—given a certain engine—than from a larger one. The smallness also reflected, perhaps, a feeling that

homebuilts ought not to compete with factory airplanes. Each had its place.

They were built by primitive methods.

There was extreme conservatism in techniques of construction, reflecting again, the golden age of twenty years before. Almost all the homebuilts were constructed of wood or of steel tubing with fabric cover, in an era in which almost all commercial airplanes were being built of sheet metal.

In the sixties most homebuilders were middle-aged or older. Whether it was the hankering of those people to return to their own heyday of the thirties, or simply the fact that older people had the time to spend, it's hard to tell. But until the sixties most of the people who showed up at the annual EAA Fly-In and Convention at Rockford, Illinois, were fairly well along in years.

In the early sixties a change began to occur. Nobody planned it, least of all the administration of the EAA, which is virtually a family business, headed since the beginning by the beloved and revered Paul Poberezny. Poberezny's loyalties have always been to the "old school" of airplane design. If there was a single event which precipitated the change, or at least symbolized it, it was the appearance on the scene of plans for a two-seat homebuilt called the T-18.

The T-18, like the Model T Ford, set the direction for everyone else. It was a clean all-metal two-seater, modern in structural conception but visibly attached to the same 1930s tradition that produced the open cockpit biplanes and parasol monoplanes still dear to homebuilders' hearts. It was a taildragger, and at first it had an open cockpit. But it had something else as well: the potential for excellent performance. The T-18 was the design an aircraft manufacturer might have built if, like the manufacturers of sports cars, he had decided there was a market for a "fun" airplane. And like a Detroit sports car, the T-18 could serve for more than sport; it provided fast transport for two people and their baggage.

Until the early seventies, the T-18 was the most popular homebuilt design; then it was eclipsed by another new phenomenon: the mass-market homebuilt. Designs began to compete with one another for the title of quickest and simplest to build, and the covers of large-circulation tinkerers' magazines like *Popular Science* and *Mechanics Illustrated* began to display plane after plane that "You Can Build Yourself!" Whereas more than a thousand people had bought plans for the T-18 over a ten-year period, several thousand could be counted on to plunk down $50 or so for a minimalist design in the first few months after its announcement.

The new designs implied a reversion to the rusticity of the earliest homebuilts, with one conspicuous exception: the Bede BD-5. The first announcements of the BD-5 represented it as a dream airplane, practically impossible to improve upon: tiny, fast, beautiful, economical, easy to build. Except for beauty, all the claims turned out to be false, but not before thousands of builders had paid millions of dollars for airplane kits and for completed airplanes which they were never to receive.

The BD-5 produced a shift in the economics of homebuilding. If the most successful marketers of earlier designs had opened eyes by bringing in one or two hundred thousand dollars, Bede opened mouths with his millions. That his scheme collapsed did not detract from its effect; all over the industry there was a new awareness of this untapped, unimagined market. The membership of the EAA also swelled rapidly thanks to Jim Bede. And it was out of the ranks of Bede's team that a new Messiah, Burt Rutan, was to come, with a design superior to the BD-5: the VariEze.

In the VariEze the homebuilding movement found the key to perfect respectability. Here was an airplane possessing impeccable technical credentials, capable of achieving its advertised performance, extremely easy and quick to build, and, what was most significant, highly innovative in both structural and aerodynamic design. Fifteen years after the introduction of the T-18, the homebuilding movement had come into its own. No longer eccentric and reactionary, it now represented a leading force in technological innovation and a significant sophistication of aircraft production.

Contemporaneously with this evolution, there had come about a change in expectations and attitudes. Builders of homebuilts no longer planned to spend their time hopping about from one grass airport to another in the Iowa summer and swapping stories about it all winter. Full

blind-flying equipment, retractable landing gear, high speeds, comfortable cabins—all the capabilities and amenities of factory airplanes —could now be found in many homebuilts. Homebuilts filed IFR (Instrument Flight Rules) flight plans between major international airports without raising an eyebrow. They flew long distances, across oceans, around the world.

Remarkably, though, this trend did not bring about the abandonment of the thirties spirit; it only broadened the EAA's spectrum. The love of the sport airplane is still there, and the majority of homebuilt projects continue to be of the rustic sort. But odd alliances have developed, as, for instance, in the Christen Eagle, the Aston Martin of sport plane kits.

An aerobatic biplane designed for world competition performance, the Eagle sells as a kit for $27,500 and requires 2,000 or more hours to build. The farmer who buys this flivver must also be a gentleman.

The explosive growth of airplane homebuilding has been possible only because of the latitude provided by the licensing regulations. An airplane which is commercially built and sold has to have a Type Certificate (TC), which it earns by passing long, costly, and repetitive tests to demonstrate its compliance with very detailed and for the most part demanding safety regulations. Once the Type Certificate has been issued, all airplanes manufactured under it must be identical to the one for

The Midget Mustang—this one belongs to Mustang Two designer, Bob Bushby—is still a popular design with home-builders, thirty years after its creation by Piper engineer, Dave Long. There have been several retractable-gear and folding-wing versions.

which the TC was issued, so far as structure and aerodynamic qualities are concerned. Any significant change must win separate approval.

Those very restrictive rules apply to what are called "Standard Aircraft." But there are some other categories, of which one is called "Experimental." It is a catchall term, applied to all airplanes that have been built from scratch but not certificated, or have been modified in such a way as to invalidate their TCs. The Experimental category contains several subheadings, such as "Research and Development" and "Exhibition," which permit operations only in limited areas and under certain conditions. But there is one subheading called "Experimental—Amateur Built" and the rules that govern it are very permissive. Once an amateur-built airplane has completed a testing period, it can be operated just as though it had a Type Certificate, with the exception that it cannot be used for commercial operations, carrying passengers or cargo for hire.

Most foreign governments are far less tolerant of amateurs creating airplanes. They either require a testing program tantamount to that used for certification, or they restrict the use of homemade airplanes. In the U.S. the private builder is allowed almost the same privileges as manufacturers —a fact which many people unacquainted with aviation would find startling. But the safety record of homebuilt planes, while far from

perfect, has justified the confidence of the author of those regulations. Homebuilts have never proven to be a public menace— witness the fact that most of the public is only dimly aware of their existence.

The motives of amateur builders are various. Some are hobbyists who move from one project to another, building an airplane as they would a boat or a dune buggy. Some are cranks intent on repealing the laws of aerodynamics. Some are inventors translating a novel model plane into a full-scale scientific breakthrough. Some have dreams of grandeur, like schoolboys who sketch swarms of fighter planes during dull classes, making machine-gun noises under their breath. Almost all are answering some obscure call, responding to a fascination with flight. But the abnormal determination and patience of all homebuilders seems to come from a passion for airplanes that goes beyond interest, beyond fascination, into the realm of religion and voodoo.

Only irrational obsession could account for the ability of so many builders to stick with the task

through its endless delays and frustrations. A few kits have lately come along that can be built more or less quickly; but thousands of homebuilts have been built in thousands of hours, over periods of years, and at unexpectedly escalating costs.

Some people think that to build an airplane is the quick and cheap way to get flying. Nothing could be farther from the truth. Building an airplane, any airplane, takes a long time and costs a lot of money. To build an airplane just to have an airplane is impractical. You have to like building. You have to like reading plans, making parts, finishing, detailing and assembling. Since it will probably take a year or several years, the work itself has to be a pleasure. For some the process is the actual goal. Many homebuilt airplanes don't fly much after they're built. Some are sold after a year or two, as builders discover that they do not have as much use for an airplane, or do not like flying as much as they supposed. And then there are builders who begin again—begin another useless airplane, because they have come to love the work itself.

It was proverbial for years that no homebuilt had ever flown so many hours as it had taken to build. Some day that will be changed by a VariEze or a Quickie or another of the new almost-instant airplanes. But it will always be true that to build an airplane you have to love airplanes, and you have to love building even more.

11

2 BD-5

Empty weight: about 500 lbs
Wingspan: 21.5 ft
Length: 13.5 ft
Seats: 1
Cruising Speed: about 200 mph
Range: about 900 mi
Landing Speed: 70 mph
Engine(s): Zenoah 3-cyl, 70 hp

It is not clear whether the BD-5 is still slightly alive or is altogether dead; either way, the question is academic. The story of the BD-5 is certainly over; it ended not with a bang but with a long series of whimpers.

The BD-5 was a milestone in the history of the homebuilding. It singlehandedly ushered in the era of mass-marketing techniques, of high performance kit airplanes with short construction times and very low prices. No other airplane before or since has promised as much as the BD-5. More than any other airplane it opened the eyes of the general public to the idea that anyone who wished could build, own and fly an airplane. People who knew nothing else about aviation in 1974 knew about the BD-5; it even appeared in *Playboy,* with a nude lady uncomfortably reclining upon a bunch of partially assembled parts. Most designers who cater to the homebuilding trade are wary of the non-aviation market —the unsophisticated customers who nourish fantasies triggered by an article in *Popular Mechanics.* Jim Bede (pronounced "beady") courted them; he welcomed them and their deposits with open arms.

Bede was always compared, even by his supporters, to a huckstering evangelist. It was obvious to the cognoscenti from the very first that the claims he made for the BD-5 could not be realized. Aeronautical professionals were aware that the then current state of the art of airplane design

allowed only so much improvement. A 32-hp airplane cruising at more than 200 mph—the claim initially made for the BD-5—was simply not possible. But there was no way the average pilot, or nonpilot, could know that. In general people are credulous about technical matters, and believe blindly in the possibility, in fact the inevitability, of continuous progress. It was an intoxication with the idea of progress that made them believe in the BD-5.

The idea behind the BD-5 was to use a two-stroke engine (because they are cheap, light, simple, powerful, and widely available) to power the smallest possible airplane for a single pilot. To get good power out of the small engine it was necessary to use high engine speed and some kind of reduction gearing. Bede decided initially upon a variable-speed V-belt arrangement, such as is

often used in drill presses, with a propeller drive shaft mounted in bearings near the top of the airplane. Such a drive system is liable to fail for reasons unfamiliar to laymen, mostly having to do with an obscure phenomenon called "torsional resonance."

Apart from the technical problems presented by the reduction drive system, there were those of the engine itself. The little Hirth snowmobile engine Bede first used simply couldn't take the strain of continuous high-power operation. What was more, the original design of the airplane was aerodynamically unsatisfactory; but that was not discovered until some time after customers started sending in $400 deposits to secure their delivery positions. At that point no airplane had yet been flown.

Bede sold thousands of kits.

In order to fit IFR instrumentation into the tiny airplane's cockpit, small-face instruments are used whenever possible. Several BD-5s had full IFR equipment. Note the annunciator lights along the panel top—the professional touch.

Money flowed in and, buoyed by what would prove to be unjustified confidence in his engineering team, he spent unstintingly on promotional campaigns to keep it coming in. When the expense of the development program exceeded the contributions from customers, he launched a new promotion. He solicited deposits for a certificated, factory-built, and at the time entirely hypothetical version of the airplane, the BD-5D. Undeterred by past failures of Bede Aircraft to deliver more than fragmentary materials kits and a few flying demonstrators (whose engines failed continually), thousands of people sent

in money for this airplane, too. Later Bede promoted yet another version of the plane, one powered by a tiny jet engine.

Many buyers took advantage of a pay-in-advance-by-installments layaway plan and believed that their deposits were being held in an escrow account. In fact the money was being spent on vain efforts to perfect the airplane. As the years of frustration and failure multiplied, insiders watched with great curiosity to see how Bede would silver-tongue the horde of angry customers who came to Oshkosh each summer prepared to flay him. He never failed to

persuade even his most outraged clients that they ought to be patient a little longer—and even pay him a little more money.

From the aerodynamic standpoint, the BD-5 was eventually perfected, although the advertised performance could never be achieved. A Japanese three-cylinder engine called a Xenoah, which could put out 70 hp, replaced the unsatisfactory Hirth, whose manufacturer had, in the meantime, gone out of business. But it was too little, too late. The Xenoah, while superior to the Hirth, was still far from ideal. The design of the airplane had become

15

quite complex and its manufacture prohibitively expensive; its reduction drive system had never been perfected; and cash had been flowing in the wrong direction for too long.

Finally Bede's money ran out altogether and creditors (who did not have the personal desire for the little plane to keep them motivated) started taking him to court. Bede Aircraft went into receivership. Attempts to shore it up with government loans failed (not surprisingly, in view of the uncertainties about the BD-5's design features); creditors had to be satisfied with the return of a fraction of their monies. The marshal locked the doors of Bede Aircraft's warehouse, and, figuratively at least, the lights started going out in garages all over the country.

Thousands of partially completed BD-5 kits—many awaiting only the delivery of the powerplant and representing not hundreds but thousands of hours of eager work—now collect dust. There have been efforts to provide the airplane with alternative engines, like a turbocharged Honda CVCC. But BD-5s that have made it into the air have had a terrible safety record, many of them crashing during early test flights, several fatally. Those BD-5s that do fly stay within gliding distance of an airport.

Despite its dismal history, the BD-5 proved that millions of dollars were waiting to be spent on dream machines. And it gave

birth to a salutary caution about advertising claims and promises. A new phrase entered the aviation vocabulary: "a Bede-type operation,"—meaning one propelled out along a limb by excessive optimism. If the BD-5 experience made life a little harder for the pitchmen of new homebuilt kits and products, it also provided the buying public with some valuable lessons. They learned that technical problems cannot always be washed away by a flood of capital; that neither the federal government nor the FAA exercise any control whatsoever over the promotion and sale of homebuilts; that the fact that something appears in print doesn't make it true; that the aviation press cannot be counted upon to reveal what it knows if it knows anything (which in this case it did), because of fear of legal reprisals; and that the safest attitude is one of extreme suspicion and cynicism.

The attitude of the press toward the BD-5 affair was particularly interesting. The trade press has an ambiguous role: on the one hand it serves as a representative of consumers and a conscience for the industry, while on the other it tries to promote aviation. The aviation press makes most of its money by selling advertising space to industry, and the industry uses that fact to exert pressure on the press. Many people in the aviation press knew years before the public that the BD-5 did not perform as promised, that its powerplant and drive system were unsatisfactory, that it

was much harder to build than was claimed, that it was unlikely to be certificated, that the so-called "short-wing" version was scarcely flyable at all, and that the factory airplanes had constant engine failures and many accidents. But the press also knew that a BD-5 cover story was good for a big jump in newsstand sales, that Bede Aircraft advertised extravagantly, and that there was danger of a lawsuit over an excessively critical article. So the press abdicated its duty of consumer protection almost entirely (except articles in one or two publications), and instead did its best to help inflate the Bede bubble, with enthusiastic photo coverage and progress reports.

The BD-5 looked so good that people assumed it was good; they assumed that something as sophisticated and potentially lethal as an airplane could not be misrepresented. They were quite wrong.

The skepticism of those who were involved in the BD-5 debacle was permanently aroused. But for the kit airplane fancier who came upon the scene since then, there are no more protections now than before. Caveat emptor!

Left
Micro Bede Aviation of San Jose, California, is the current hope of those who are encumbered with unfinished BD-5 kits. Micro Bede stretches the fuselage five inches, installs a turbocharged Honda Civic engine, a redesigned drive train, and a belly air scoop for the water radiator. Time will tell.

3

RUTAN'S RENEGADES

Opposite
Quickie
Empty weight: 260 lbs
Wingspan: 16 ft
Length: 17.3 ft
Seats: 1
Cruising Speed: 120 mph
Range: 570 mi
Landing Speed: 53 mph
Engine(s): Onan 18 hp

Overleaf
VariEze
Empty weight: 530 lbs
Wingspan: 22.3 ft
Length: 16.8 ft
Seats: 2
Cruising Speed:
165–201 mph
Range: 1100 mi
Landing Speed: 60 mph
Engine(s): VW or
65–100 hp Continental

Canard is French for duck. It is also English for an apocryphal story. Finally, and least commonly, it is the name of a certain kind of airplane whose wings are located near the back end of the fuselage, like those of a duck in flight, and whose nose, sprouting small supplemental wings, sticks out a long way ahead of the main wing. Some of the earliest airplane designs were canards—the Wright Flyer, for example—but the "tail-first" configuration was quickly almost entirely abandoned. A few attempts to resurrect it were notable failures. The canard Grumman Ascender, for instance, came to be known as the "Ass-Ender" after ending several test flights by falling out of the sky backwards. To most designers the concept seemed intuitively wrong, like putting the feathers on the front end of an arrow.

In the sixties, however, a few delta-winged fighters started to use small canard surfaces (the name applies not only to the configuration as a whole but also to the forward wings) for additional lift at low speeds. But it took an ingenious young aeronautical engineer from California to bring the canard back into wide acceptance.

Burt Rutan had left a job as a flight test engineer with the Air Force to work for Jim Bede on the BD-5. While with Bede—in his spare time—Rutan built the first full-scale version of a canard airplane from a design he had developed using radio-controlled model tests while still in college. The VariViggen—he named it in

honor of a Swedish canard jet fighter, the Viggen—was an all-wood, tandem two-seater with retractable landing gear and a pusher Lycoming engine of 150 horsepower. As expected, the VariViggen proved highly maneuverable in low-speed flight; it could not stall or spin, so it could be flown with abandon right down to its minimum speed. The minimum speed was fixed by the stalling of the canard; the rear wing never stalled. A conven-

tional airplane loses a large portion of its lift when its wing stalls, and it must dive to recover flying speed. When the VariViggen's canard stalled, the plane made only a mild nodding motion and kept right on flying.

While safety was the single most important advantage of the canard arrangement, it had plenty of others, too. For one thing, the cabin could be put between the wing and the canard, an engine attached to either end or to both, and no other structure was required. Conventional airplanes, on the other hand, had to have a long tail cone to put the tail at a distance from the wing, but to maintain proper balance the cone could not be entirely filled with payload. A canard airplane of a given cabin size was also smaller and lighter than its conventional counterpart.

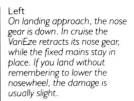
*Left
On landing approach, the nose gear is down. In cruise the VariEze retracts its nose gear, while the fixed mains stay in place. If you land without remembering to lower the nosewheel, the damage is usually slight.*

The impeccably detailed cockpit of Mike Melville's VariViggen betrays a military inspiration. Less efficient aerodynamically than Rutan's later airplanes, the Viggen nevertheless demonstrated the practicality of his canard design concept.

Like a camel, the VariEze kneels to disembark its occupants. If it did not, with the cockpit empty it would keel over backward—a design problem with all rear-engine airplanes.

And finally, the canard was simple, with no high-lift devices like landing flaps. Instead it extracted a great deal of lift from the canard surface, eliminating the drag and down-load of the conventional horizontal tail.

In 1975 Rutan left Bede and returned to his native California to apply his ideas to an airplane which began, in his head, as a canard version of the BD-5. He found that he could fit two people, rather than one, into the package. He began with a metal prototype pieced together of BD-5 parts. Deciding that was too heavy, he set out to build the definitive version of the new airplane out of plastic foam and epoxy-fiberglass composites.

The first prototype, which had no ailerons and depended on the canard's moveable surfaces for both roll and pitch control, was barely manageable in the air, and its converted Volkswagen engine turned out to be unreliable. After much experimenting with ways to improve the control system without complicating it, Rutan finally added inboard ailerons on the wing. The handling problems disappeared. He built a second prototype to handle slightly larger engines—up to the 100-hp Continental 0–200—and in 1977 began selling plans to homebuilders.

The VariEze, as he called the new airplane, was a runaway success. (The name puns on VariViggen and "very easy;" it was supposed to be very easy to build and fly.) Rutan sold several thousand sets

of plans and complete or partial kits; hundreds of VariEzes are already, or soon will be, flying.

The EZ, as it is called for short, is a highly efficient airplane attaining cruising speeds of nearly 200 mph (if it's built properly and kept light) with a 100-hp engine. It gets 30 to 40 miles per gallon. It takes between 600 and 1,200 man-hours to build, if most of the available prefab parts, like landing gear, canopy, cowling, and so on, are bought from suppliers. Although the construction system is unfamiliar to most people, builders find it easy to learn, thanks to the step-by-step illustrated instructions of the huge construction manual. The principal problem is epoxy allergy, which some people experience after being exposed to composites for a while. Some have suffered so badly they have had to give up and sell their par-

tially-completed projects. The risk of allergic reactions can be reduced, if not eliminated, by good ventilation and the use of protective clothing and skin coatings.

The EZ is such an extraordinary little airplane that most owners try to get more out of it than Rutan intended. Planned as a good-weather daytime-flying airplane, EZs end up loaded with equipment for night and instrument flying. Planned for spartan furnishings and a smooth but minimal fill and finish, they end up with luxury cockpits and extra pounds of fillers and paint. The typical VariEze is far overweight, and the problem is exacerbated by widespread second-guessing of Rutan's engine specifications. Pressure from builders forced Rutan to approve a 235-cu. in. Lycoming installation. Some builders have gone to an unapproved 290-cu. in.,

125-hp engine. Now Rutan has bowed to the tide, and is developing a slightly larger version of the airplane with a bigger, repositioned wing to handle the heavier engines. The reason is simple: people are going to use larger engines anyway, and the 235-cu. in. Lycoming engine is actually less expensive than the no-longer-in-production 200-cu. in. Continental.

Even though the EZ seemed, when it was being designed, to cater to the market Jim Bede had tapped —the market for a high performance airplane that would be cheap and easy to build—it turned out to be fairly expensive. One or two builders claim to have built EZs for around $5,000 (the precision of their accounting is unverified), but most find that by the time they get airborne they have around $10,000 invested in their airplane. A good portion of that is the engine.

After the EZ there was still a market for a truly minimal airplane which emphasized cost first, ease of construction second, and performance third. Tom Jewett and Gene Sheehan, both friends of Rutan and refugees from the Jim Bede organization, decided to build an airplane around an industrial motor which

In allusion to its ambiguous shape, Rutan christened his own original VariEze "Glass Backwards."

23

had the toughness and reliability of an aircraft engine but could be purchased for a few hundred dollars. They settled on a two-cylinder air-cooled Onan motor of 16 horsepower. The only rub was that it weighed 80 pounds.

Jewett and Sheehan hired Rutan to design an airframe that could make the most of their unlikely engine. The airplane he came up with, if it proved nothing else, showed that Rutan was not a man of only one design.

To keep weight and complexity down, Rutan hit upon the idea of using the canard surface as a main landing gear. This required putting the weight well forward. The result was the Quickie, with front engine, tandem-wing design, and a long tail cone which sacrificed the VariEze's advantage of structural compactness. To minimize the weight and surface area of the fuselage, Rutan undercut it, giving the airplane a curious shape, like that of a leaping fish or a copulating dragonfly. Odd as it looks, the Quickie is in most respects an aerodynamic success. It can hit a speed of nearly 130 mph, cruise 100 miles on a gallon of gas, and maneuver in low-speed flight

with even greater facility and security than the VariEze. It is also phenomenally easy to build: 400 man-hours is the target time. Using a high-compression version of the Onan engine which develops 18 hp, the Quickie climbs 400 fpm—not a terrific rate of climb, but an adequate one, especially since precise speed control is not necessary to achieve it. In fact, with the stick held all the way back—a condition in which a conventional airplane would be going down, not up—the Quickie still climbs at 150 fpm.

In a few years Rutan's designs

have come to be the best known in the field; eventually they will almost certainly be the most numerous as well. But what is most remarkable about his work is not its commercial success or its popularity but the originality and good judgment he has consistently shown. There are two ways of designing. One consists of studying other versions of the device one wants to build and then designing a new one with certain modifications or improvements. The other—sometimes called "the systems approach"—consists of analyzing a problem from scratch, using intuition as much as reason, and attacking it as though it had never been dealt with by any other designer before. Most engineers stick to the first method. It is safe, dependable, and does not require extreme talent —and there is a certain pleasure and satisfaction in producing a more elegant version of the tried-

and-true formula. Rutan uses the second method. He does not even copy himself. His originality is so complete, and some of his inspirations so bizarre, that one is at a loss to criticize them. There are no points of reference or comparison. And one by one they have proved themselves. Perhaps a conventional airplane could be designed to equal the performance of the VariEze, but no one has actually managed to do it.

The very least one could say is that Rutan has singlehandedly given aviation faith in a new kind of design and, by so doing, has rekindled the ambitions and imagination of all designers. In that sense he will have been responsible in part for a generation of new designs.

Rutan's most recent project, following the Quickie, is not for

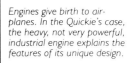

homebuilders but for commercial manufacture. Dubbed the Defiant, the new airplane is a four-seat twin, similar in general arrangement to the VariEze but with engines at both ends of the fuselage. As was the case with the VariEze, the first prototype proved the concept. A second prototype is being built to establish the design of the production airplane and the molds that will be used to make it.

The Defiant outperforms the conventional, factory-built light twins with which it will compete, while at the same time providing a much greater margin of twin-engine safety. If it had been the invention of Cessna, Piper, Beech or Grumman it would already have obliterated its competition. But it is the product of a renegade outsider, and the financial obstacles to getting an airplane into production and keeping it there are fantastic. They may be greater than the force of a merely superior product. So for a long time to come, the only way to own a Rutan—the latest and best in airplane design—is to build one yourself.

Engines give birth to airplanes. In the Quickie's case, the heavy, not very powerful, industrial engine explains the features of its unique design.

4 | THORP
T-18

Empty weight: 900 lbs
Wingspan: 20.8 ft
Length: 18.9 ft
Seats: 2
Cruising Speed:
165–185 mph
Range: 580 mi
Landing Speed: 60 mph
Engine(s): Lycoming
125–180 hp

Twenty years ago, when the T-18 was about to be designed, home-built airplanes were not, as they are today, more advanced than comparable classes of factory-built airplanes. Most of the factories had turned to sheet aluminum construction. The homebuilders, however, still worked in wood, tube, and fabric in the mistaken belief that, because these methods were familiar and time-honored, they were necessarily better. Even if they had wanted to build an all-metal airplane, there were few all-metal designs to be found. John Thorp, a professional aeronautical engineer, changed all that.

Thorp worked for Lockheed until the end of World War II, then went off on his own to develop a two-seat light airplane, the Sky Skooter. He certificated it but, as often happens in the uncertain aviation market, the airplane didn't stay in production for long. Drawing from the Skooter, Thorp later created the basic format for the Piper Cherokee series. In developing this series, he aimed for a minimum of parts and simplicity of assembly—essential elements in the design of a modern airplane, factory- or home-built. These same considerations were incorporated into his plans for the T-18, his next project.

Thorp remembers clearly what initially impelled him to design a homebuilt. He had been in a salvage yard where he saw a lot of Lycoming 0-290-G engines for sale for a few hundred dollars each. To him, these inexpensive engines were a sign. He knew, as does every airplane designer, that engines lead the way to airplanes. These Lycoming engines not only led Thorp to the T-18, they practically forced him to follow.

In light airplane design there are specific points—the landing gear, the cockpit canopy and windshield, the engine cowling, the fuselage contour behind the cabin (where the sides begin to converge), the wing root and wing-fuselage attachment, to name a few of the most obvious—where cheapness and simplicity must be blended with performance and beauty in certain proportions, and in a certain order, to produce an aerodynamically acceptable result. Thorp designed a tight two-seater with a tailwheel; he left off the cowling and canopy altogether. He solved the fuselage convergence problem by choosing a polygonal cross section for the fuselage and providing it with fair curves lengthwise—the way the air flows. He moved the wing joints outboard, where they would carry a smaller bending load and be lighter; and, for convenient trailering and storage, he made the wing easily removable.

His plans called for a "matched-hole" system of construction in which rivet holes are drilled in most parts in advance, while each part is still a flat blank of aluminum lying on the table. After being bent and formed, the parts could then be riveted directly together. No assembly jigs would be necessary because, with the holes precisely located in the first place, the symmetry and straightness of the finished airplane would be assured.

Thorp saw this introduction of an all-metal two-seater to the home-built market as a rather ambitious scheme. He felt he was doing it principally for his own satisfaction, that probably no more than fifty people would be willing to purchase the kit and build the airplane. But he was wrong. The T-18 was an instant success; the plans have sold at a steady rate of from five to ten sets a month for the last seventeen years. And, like so many other well-conceived designs before and since, the T-18 fired the ambitions of its builders to surpass the limits of the original design.

The very first builder, Bill Warwick, substituted a 180-hp Lycoming engine for the 125-hp one around which the plane had been designed. Because the big engine significantly increased the airplane's speed, the open cockpit was filled with a constantly raging tempest, so a canopy was made to enclose it. A cowling followed soon after, then wheel pants. While all these modifications eventually became standard accessories, some builders went even further. One converted the airplane to a tricycle retractable landing gear; another stretched the fuselage to accommodate two children behind the front seat; still another developed a trailerable folding-wing version. The T-18 was becoming quite a sophisticated little airplane, but it wasn't until Don Taylor, a retired Air Force meteorologist from Hemet, California, began entertaining his own extravagant fantasies that the airplane surpassed even John Thorp's wildest expectations.

Don Taylor wanted to fly his T-18 around the world. In preparation for this daring jaunt he equipped his T-18 to carry about 150 gallons of fuel, crammed it with radio and survival equipment, and left just a tiny space in which he could sit. In 1974 he made his first attempt to circle the globe. Unforeseen delays in Japan held him up until bad weather closed in on the Aleutians and he was forced to abort his trip. He shipped the airplane home, made further modifications, and tried again in 1976. This time he succeeded, overcoming the obstacles of distance, weather, fatigue, bureaucracy, and the churlish attitude of the U.S. Navy toward his use of some of their Pacific island landing fields.

Thorp had been against the trip from the beginning, thinking the over-water legs too dangerous and the airplane's systems not sufficiently redundant for that kind of flying. But Thorp usually tended to underestimate his airplane, as well as the ingenuity and determination of its builders. They continue to surprise him. One, Clive Canning, flew his T-18 from his native Australia to England, being forced down along the way by Syrian Migs. Another builder recently set some kind of record by finishing a T-18 after working on it for sixteen years!

The T-18 is today a classic home-

Many T-18s have been built with full blind-flying instrumentation, and a few with turbochargers for high altitude flight.

29

built. With a 180-hp Lycoming engine, it cruises at nearly 200 mph and easily climbs 2,000 feet per minute. It has a fairly limited range, about 500 miles, because the airplane's 28-gallon, fuselage-mounted fuel tank wasn't intended to accommodate such a big engine. The design is somewhat old-fashioned, making the airplane, in Thorp's own words, a "modern antique." It is, he says, what would have been built in the thirties if those inexpensive little engines had been available then. Nevertheless, there are few newer designs that surpass it. A great many prefabricated parts are available from the usual sources, and partially completed T-18s often turn up for sale. They are surprisingly expensive; but the day of the $3,000 airplane is long gone. A typical T-18 built today costs around $10,000. For that tidy sum you are committing yourself to at least a couple of

years of work on an elegantly simply all-metal structure, a rather cramped but hot little ship which, even when sitting on the ground, appears to be leaping forward—much like the tiger for which it was named.

In the air, the T-18 responds briskly to the controls, which get perceptibly heavier with increasing speed. The airplane, despite its rather small vertical tail and large propeller, is quite stable and recovers well from a spin; one pilot did eighteen turns and recovered immediately. The rudder is very powerful, but the stabilator is marginal under some circumstances; with a forward loading and flaps at low speed, the stabilator can actually stall and produce a disconcerting pitch down. The quality of the stall varies from one airplane to another. It can be tailored to be as gentle as a Cub's or, left to itself, with

slight errors of wing twist and leading edge radius, it can be quite violent.

The T-18's wingspan is short; in flight the tip looks as though it's right under your elbow. With a large engine and IFR equipment the wing loading gets quite high and the landing speed rises accordingly; but once you get the hang of coming in rather hot and flat and keeping an alert foot on the rudder after touchdown, the airplane is not unduly hard to land. The landing gear is very stiff, its only shock absorption coming from the tires and the rather meager flexibility of the steel-tube legs. But despite the stiffness, many owners fly their T-18s into and out of grass and dirt fields and manage well enough.

John Thorp's T-18 is a proven success; so far, around 300 have flown—a few well over a thousand hours. Its success is easy to understand. The T-18 is a perfect compromise between the performance and utility of a factory airplane and the rusticity and compactness of a homebuilt. Though the basic design is old, the T-18 is still one of the most practical, usable airplanes one can build. It looks strange enough for people to stop and stare, but it behaves enough like a commercial product to spare its owner from ever having to apologize for its being a homebuilt.

The fundamental idea of the T-18's aerodynamics is to stagger individual objects: the wheel pants begin to taper as the wing section begins to expand, the fuselage and cockpit canopy expand as the wing section tapers, and so on. The effect is to minimize the total displacement of air required by the passage of the airplane. The only portion of the airplane in which Thorp allowed himself the luxury of sensuous shape was the cowling.

5

POLEN SPECIAL

Empty weight: 1035 lbs
Wingspan: 21.5 ft
Length: 19.5 ft
Seats: 1
Cruising Speed: 312 mph
Range: 1400 mi
Landing Speed: 65 mph
Engine(s): turbocharged 200 hp Lycoming

This is the last word in home-builts. It is a World War II fighter in every respect except size. It not only has the performance of a fighter, but also has much the same systems (except the armament) and provides the same cockpit ambience. It is a single-seater with power out of proportion to its size, and as such belongs in a different category from the kinds of airplanes that are intended for two, for week-end travel, or for the amusement of low-time pilots. Dennis Polen, who designed and built the Special, is an aviation professional who is qualified to fly cabin-class twins and jets. The fact that his airplane is so small and so powerful that incautious use of the engine can send it out of control at low speeds does not disturb him; the same thing was true of the F4U and the P-51. The fact that it operates routinely under IFR restrictions above 18,000 feet, and that the pilot must make constant use of oxygen, does not disturb him either; a good pilot can handle such features. In fact, nothing about Polen's airplane disturbs him. He, like everyone else who sees the airplane and fantasizes about possessing something like it, is turned on by the exotic, challenging, dangerous, extremely virile quality that it exudes. The pilot of an airplane like this enjoys the awe it inspires in others.

The Polen Special is not the only airplane in its category. Home-builts of not only high performance, but extremely high performance, are becoming more and more numerous. John Thomp-

son, a Tucson homebuilder of the fanatical sort that works actively on two airplanes at once, has a 310-hp engine in a drastically modified Bushby Mustang II and reports cruising speeds in the 300-mph range. And plans are available for the two-seat tandem Brokaw-Jones 520, alias the Brokaw Bullet, which achieves nearly similar performance. The trick is simple enough: a small, clean airframe with retractable landing gear is equipped with a large turbo-charged engine. Turbocharging is the key to speed. The maximum cruising speed of a turbocharged airplane is reached at 20,000 to 22,000 feet, where the thin air exerts a greatly diminished drag on the airframe, while an exhaust-driven air compressor keeps the engine power up at the same 75-percent-of-maximum level that it would normally have at 7,000 feet or so. The result is generally

a gain of 30 or 40 mph over what would be possible without the turbocharger. For this speed and altitude one pays a price in weight and complexity, and in the legal restrictions that are imposed on flights above 18,000 feet; but that price itself conveys on the payer the prestige of a professional.

The Polen Special is at the moment the fastest of them all, though it is not so hot an airplane as the Brokaw Bullet. The Bullet is a very heavy two-seater, but its wing area is smaller than that of the Polen, and its span less. Consequently the Brokaw takes off, approaches, and lands at much higher speeds, but has a lower rate of climb and demands more of its pilot. Polen's airplane, which looks a great deal like a scaled-up Midget Mustang with a Twin Comanche cowling, represents a rather wise decision to resist the temptation of trading wing area wholesale for power. The wing is lightly loaded, the proportions of the airplane are elegantly conventional, and the reward is a phenomenal rate of climb (nearly 4,000 fpm) with only a very small penalty in top-end cruise. A low wing loading (16 lbs/sq ft) and a landing flap bring the minimum speed to a manageably low 65 mph. The Brokaw, by comparison, with a wing loading of nearly 38 lbs/sq ft, stalls at 83 mph. The benefit of the low stalling speed is that the Polen can land comfortably on conventional—that is, tailwheel—gear,

and dispense with the weight and space requirements of a retractable nosewheel.

The engine is a fuel-injected, turbocharged 200-hp Lycoming. It maintains that power to 18,000 feet, where the airplane attains its maximum level speed of 325 mph. At 22,000 feet, burning ten gallons of fuel an hour, the engine puts out 150 hp and yields a best cruising speed of 312 mph. Given those altitudes, and the possibility of enroute encounters with rough weather and icing during climb-out, the wisdom of the light wing and span loading and the very

high rate of climb is apparent. The high cruising altitudes are well within the practical reach of the airplane even for short flights; exposure to layers of adverse weather is short.

In addition to fabulous climb and cruise performance, the Polen Special is strong and responsive, fully aerobatic, with a 200 deg/sec roll rate. In a mouth-watering article in *Sport Aviation* that must have left a lot of pilots dehydrated, Polen wrote about climbing to 18,000 feet in seven minutes from starting the takeoff roll for a 160-mile flight, getting a 335-mph

The Brokaw Bullet is a design motivated by the same dreams and goals as the Polen Special—but less elegantly executed.

groundspeed readout from an Air Traffic Controller, and then, on being cleared down to 10,000, rolling over on his back into a dive that brought the groundspeed up to over 400 mph and disposed of the 8,000 feet of altitude in little more than a minute!

That kind of flying puts heavy demands on both pilot and airplane. Very high airspeeds bring with them the danger of flutter—destructive vibration of the airframe, resulting from a complex combination of air forces and structural flexibility—which can tear an airplane apart in a fraction of a second. Consequently, an airplane designed to fly at five-eighths of the speed of sound—as is the Polen Special—must be carefully engineered, scrupulously built and balanced, and cautiously tested at gradually increasing speeds. The pilot is subject to much the same demands as those made on pilots of turboprops and jets, which are normally crewed by two, one flying the airplane and the other taking care of navigation, radio work, and systems checks. Polen's article consisted largely of long checklists, which conveyed some idea of the complexity of the airplane; but it did not dwell on the difficulties that can arise while flying at high speed through bad weather: accepting clearances, planning ahead, keeping up with navigational tasks that are tossed at a pilot, one after another, as fast as he can deal with them. A reasonably skillful pleasure pilot could certainly handle the Polen at low altitudes and in good weather,

given its low wing loading and conventional controls. In its element, however, at high altitudes and high speeds, it needs a professional. For that reason—and also, one easily imagines, because drawing and selling useable plans is an immense and not very rewarding task, and perhaps also because part of the satisfaction of having such an airplane is in its being one of a kind—Polen refuses to heed the pleas of hundreds of supplicants who want Polen Specials of their own. He has, in fact, acquired a reputation for an excessive, almost surly love of privacy, which isn't surprising; people don't realize how irritating it can be to have to say no again and again. Polen is a king in his own small kingdom, and there is no reason why he would want to start mass producing crowns.

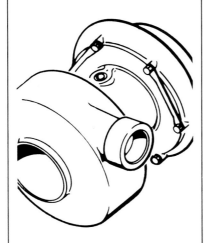

What Polen has done, however, can be and will be surpassed.

Several projects are under way that will outperform even the Polen Special. There is a tiny Unlimited racer, powered by two turbocharged Mazda engines, which may be faster (but will certainly be less reliable); it is now in the early stages of flight testing. Somebody is building a little jet in a city apartment. A father and son in Tucson are working up a tandem two-seater powered by a 630-hp Lycoming turbine that should cruise at about 400 mph. For all these and similar projects to surpass one another in succession is perfectly natural; each king sets the measure of the challenger who will depose him. There is no reason why a homebuilt could not, for that matter, attain supersonic speed or climb to 100,000 feet.

The Polen Special nevertheless represents a kind of landmark. Jets and turboprops are extremely costly and often very inefficient. Enormous engines do not always sit well in small airplanes, and all the normal values —comfort, safety, good flying qualities, versatility—end up being sacrificed on the altar of a single performance point. Various airplanes may surpass the Polen in various ways, but they will be hard put to equal its happy blending of performance, efficiency and beauty, or the nostalgic elegance that animates its form, part racer and part fighter, even as it rests on the ground. It is an extremely judicious design, radical in purpose, yet conservative and understated in execution. It will remain a classic.

The addition of a turbocharger greatly enhances the high altitude performance of virtually any airplane.

6 | MELMOTH

Empty weight: 1500 lbs
Wingspan: 23 ft
Length: 21.5 ft
Seats: 3
Cruising Speed: 230 mph
Range: 3,400 mi
Landing Speed: 70 mph
Engine(s): turbocharged
Continental, 210 hp

I hope that it will not be regarded as a breach of good manners that I include my own airplane in this collection. *Melmoth* is unusual. It represents an extreme point on the spectrum of homebuilts, and I imagine that another writer, even without my predilection for the airplane, might have included it; so I take the liberty of doing so myself.

Melmoth—the name comes from a nineteenth-century novel called *Melmoth the Wanderer*—is the crystallization of a dream of international travel. I wanted an airplane that could carry two people and ample baggage, in comfort, anywhere in the world. If one does not want to depend on the fickle mercies of the U.S. Navy for the use of island landing spots in the Pacific, the shortest range that such an airplane can have is about 2,500 miles without reserves or, say 3,000 miles with. Comfort requires a spacious cabin, plenty of leg room, reclining seats, a low noise level and a high cruising speed. It is apparent "from inspection," as geometers say, that a two-seater filling the bill could be built with an engine of about 200 hp, and a fuel capacity of around 150 gallons.

When I first thought about *Melmoth*, I knew nothing about aeronautical design or about engineering; I had majored in English in college and never went beyond second-year algebra in high school. I knew nothing about mechanical design either, having never tinkered with cars; nor did I understand the first thing about electricity or hydraulics. It puzzles me in retrospect that all this ignorance did not deter me. But I assumed that what I did not know I could learn (I underestimated the vastness of what I did not know, as the ignorant always do) and so I began studying, sketching, dreaming.

The process was long. Its very earliest stirrings occurred in 1963, when I was still thinking about a single-seater. It was not until 1968 that the design for a rather size-

able two-seater began to take shape in my mind; in 1969 I began making parts for it, and in 1973 it finally flew.

Fond of airplanes from childhood, I designed *Melmoth* according to my own aesthetic criteria, letting my imagination supply the specific outlines which calculation can only roughly sketch. Uncertain about structural design, I pored over the few pages of an engineering manual that I was able to understand, rounding off all figures upward and throwing in extra stiffeners, extra skin thickness, extra rivets everywhere to buttress my insecurity. I consulted almost daily with John Thorp, the designer of the T-18, about those basic principles and decisions which books fail to explain and about details which bewildered me, but which experience and instinct permitted him to dispose of easily.

The final configuration is never entirely final. There are always improvements, second thoughts, simplifications, complications; the airplane you see is a manuscript hundreds of times rewritten.

Always believing that once a few unexpected complications and difficulties were cleared up I would be able to finish the airplane in six months, I worked a laborer's hours beneath a grape arbor in my Los Angeles backyard. It took four years. Prevented by inexperience from finding ingeniously simple solutions to design problems, I solved them all by accretion, adding parts and systems to offset the inadequacies of other parts and systems, creating an airplane that was monstrous, ominous in its superfluous complexity. Short

on money and intimidated by the accommodations they would have required, I never tried to adapt existing parts—landing gears, cockpit canopies, fairings, cowlings—to my design, but instead designed and built every part from scratch.

During the four years of building *Melmoth,* I tormented myself with a whole arsenal of worries: was I avoiding Real Life by burying myself in this project? Was I a crank? What if it flew badly, was unstable, uncomfortable, uncontrollable, impossible to live with?

What if, in my ignorance, I was missing some essential point about design, like those New Guinea islanders who build little mock runways on mountain tops in hopes of luring airliners to bring supplies to them? Were my friends laughing at me behind my back?

On September 6, 1973, I learned that I need not have worried. *Melmoth* flew.

It took me years to get out of the habit of being with the airplane every day. After I began flying it I

41

began changing it, correcting mistakes and oversights, improving, streamlining, decorating. I changed the configuration of the tail surfaces, added landing gear doors and rear windows, filled the surface irregularities, painted the airplane. Over the next five years I obtained and installed elaborate instrumentation and radio equipment, until every empty space was filled and new spaces had to be made.

In 1974 I flew from Los Angeles to Guatemala with Nancy Salter, who had lived with me during the whole process of building the plane. The next year we took the plane—or it took us—to Europe in an 11-hour, 2,000 mile nonstop hop from Newfoundland to Ireland. In 1976 we flew from Alaska to Japan, 2,650 miles in under 15 hours. As I write this, *Melmoth* has flown more than 1,400 hours, covered more than a quarter of a million miles, and has substantially fulfilled the seminal dream. I have ceased going to the airport every day to work on some detail or other, or just to see the plane. I still indulge in modifications and improvements—the latest was the addition of a turbocharger— but I plan them carefully, make all the parts in advance, and try to spend as little time as I can grovelling under the plane and poking in its innards. I try, instead, to fly it often.

Before turbocharging, *Melmoth* would climb at 1,800 fpm after takeoff, with a couple of people and three or four hours' fuel aboard. At sea level it would do about 210 mph, burning around ten gallons an hour. Turbocharging changes the performance spectrum, producing startling "maximum" speeds which one rarely has any occasion to experience. The maximum cruising speed, on 10.5 gallons an hour, is 235 mph or so at 22,000 feet; but more normally I set up a 210-mph cruise at around 16,000 feet on 10 gallons an hour, turn on the autopilot, put on the stereo headphones, and take it easy. *Melmoth* is well equipped, with full blind-flying instrumentation and radios and a lot of amusing little gadgets, like a system that automatically switches fuel tanks, from right to left and back, every five minutes, to keep the fuel quantities in the wings balanced. Its landing gear and flaps are hydraulically operated; and there are spoiler speed brakes in the wings, also hydraulically actuated, which makes it possible to descend at more than 4,000 fpm at 200 mph while still running the engine hard enough to keep it warm.

The four fuel tanks hold, in all, 154 gallons of fuel, sufficient for a nonstop flight of 3,400 miles. Normally, of course, I carry only 30 or 40 gallons; the plane is far more fun to fly when light. A jump seat in the baggage compartment can accommodate a third occupant on short trips.

Very surprisingly, *Melmoth* succeeded beyond my expectations in most respects. It has even managed to operate successfully into and out of dirt strips in Mexico and Central America, and, although designed for a maximum gross weight of 2,350 pounds, it took off from Japan for the U.S. at a weight of 2,950.

In two areas I fell far short of my hopes. One was noise reduction inside the cabin, at which I failed completely; the other was empty weight. At 1,500 pounds *Melmoth*

is 500 pounds above my rather naive design estimates, and 225 above the weight at which, in stripped-down condition, it originally flew. Overweight is the common blight of amateur-designed airplanes, but not many weigh so much as this.

Melmoth has meant even more to me than I thought it would when I was building it. While building I insulated myself against too-high expectations, for fear of disappointment. I tend to take the plane for granted now. But from time to time, I break out of the top of the clouds on an IFR climb and, looking back over my shoulder out the rear windows, I seem to see the airplane sliding along close to the clouds at a terrific clip. At those moments I get a sense of its speed and beauty, as though I were racing along beside it, looking on. It is—absent the novels and poems that I never wrote, the painting I did not paint, absent all the things I once thought important and imagined that I was destined to do—the only beautiful thing I have ever created.

An ambitious airplane merits ambitious instrumentation; Melmoth's is like that of a small airliner.

7

DYKE DELTA

Empty weight: 960 lbs
Wingspan: 22 ft
Length: 18.5 ft
Seats: 4
Cruising Speed: 170 mph
Range: 725 mi
Landing Speed: 70 mph
Engine(s): Lycoming
125–180 hp

One thing that tends to make factory airplanes pretty dull is that they are designed to satisfy the needs of the greatest number of prospective purchasers. Hence, the process of their design is fundamentally one of compromise; it has a large statistical element. Homebuilders, on the other hand, have only themselves to please—especially those homebuilders who design their own airplanes.

John Dyke has his own ideas about what an airplane ought to look like, and what it ought to do. When he started designing his Delta in the early sixties, he wanted a roadable (one that could also operate as an automobile) four-seater; above all else, though he didn't exactly say so, it seems he wanted a delta-shaped airplane. Now if Dyke had been an ordinary airplane purchaser he probably would have searched high and low for a delta-winged Piper, Cessna or Beech and come up with nothing. But he wasn't and, hacksaw in hand, he set to work to build his own.

He started with model tests, first with crude, free-flight models, then with a larger version that flew in a force-measuring rig on the roof of his car which measured the basic aerodynamic parameters of his unusually shaped airplane. Finally, he built and flew a full-scale airplane, though it was destroyed in a fire in 1964. Applying what he had learned from his first prototype, he designed and built a second, the JD-2. Dyke now sells plans of the JD-2 to homebuilders, and copies are being put together around the world.

The Dyke is, to say the least, an unusual airplane. Not only does it look strange, it also has unusual flying qualities, and it is made of unusual materials. You would think that its designer had never been influenced by any other airplane—except, perhaps, the F-102 Delta Dagger.

The structure uses a framework of steel tubing to support strips of stainless steel channel, to which skins of fiberglass sheet, molded by the builder, are held with explosive rivets. The choice of fiberglass skins was necessitated by the inconvenience of skinning the odd-shaped airplane with aluminum sheets that come only in standard sizes. The substructure is joined together by nickel-silver brazing. A description of the Dyke structure is remarkable in its omission of any of the normal vocabulary of airplane structures —aluminum, rivet, weld, glue, wood, and so on. Nevertheless, the specifications call out an empty weight of 960 pounds—about the same as that of a T-18, and about equal to the airplane's payload. Of course, to paraphrase the caveat that always accompanies EPA figures in auto advertisements, *your* Dyke Delta's weight may vary.

Delta-wing aircraft were developed to operate in the transsonic and supersonic speed range (700–1000 mph); their low-speed lift is developed by a mechanism somewhat different from that of conventional wings, and at the price of considerable low-speed drag. For military aircraft equipped with big jet engines and designed to fly faster than sound, excessive low-speed drag is unimportant; they can overcome it with a surplus of engine power. For light aircraft, however, the situation is different. The surplus power available is comparatively small, and low-speed drag seriously penalizes both climb performance and the ability to take off and land safely on short runways.

At low speeds the Dyke Delta pays a penalty for its shape. However, that penalty does not cripple it completely, because the Dyke Delta is not really a delta. A delta wing, in the limits of normal usage, is one that is triangular in shape, with a high angle of leading edge sweep, an essentially straight trailing edge, and a short span compared with its centerline chord. The JD-2 is really a low aspect ratio flying wing. Its predecessor, the JD-1, was a true delta, with 57 degrees of outer panel sweep to the JD-2's 31 degrees, an 18-foot span to the -2's 22 feet, and pointed wingtips. In recontouring the wing, Dyke nearly doubled the area of the outer panels, raised the aspect ratio from just under 2 to 2.71, and clearly abandoned every vestige of the delta but the name.

The JD-2 has a span loading and lift distribution much like those of a conventional airplane; it differs from a conventional airplane in having a very low wing loading—

less than 10 lbs/sq ft. Now, a conventional airplane with that low a wing loading, like a Piper Cub, would approach and land at 50 mph; but the Dyke, because it makes extremely inefficient use of its wing area, approaches at 90, touches down at 60, and cannot operate comfortably out of short or obstructed airstrips. On the other hand, the Dyke is quite clean, since its generous wing area practically absorbs all the other drag-producing areas—fuselage, tail surfaces, and surface intersections—found on conventional airplanes. The result is an airplane featuring good cruising performance—180 mph with a 180-hp engine—but poor short-field and low-speed characteristics.

This low-end slump is aggravated

by weight: with only the pilot and a couple of hours' fuel aboard, the Dyke climbs at 1,800 fpm at 110 mph, but its rate of climb diminishes by 200 fpm with each 100 pounds added to the useful load. Thus, while it is possible for four people to get into the cabin (which is reminiscent of one of those cardboard boxes in which, as little children, we conducted our imaginary bombing missions), they would be well advised not to go travelling on a hot day, or at high elevations.

The Dyke's seating position is unusual, with the pilot placed in solitary splendor on the centerline, and two or three passengers side by side behind him. The floor is flat, and everyone sits on cushions on the floor, creating a

Middle Eastern ambience. The triangular disposition of the occupants solves the problem of the limited center of gravity (CG) range available with a tailless airplane. In the Dyke, the pilot is figured into the empty CG calculation, and all the other disposable loads—fuel, passengers, baggage (which is carried in a well in one wing root)—are then clustered around the CG, which always remains virtually unmoved despite a wide variety of payloads.

Tailless airplanes have difficulty getting their noses up at low speeds, for instance during takeoff. The Dyke however sits on the runway at a 7-degree nose-up angle, so that it is already at the takeoff attitude just sitting there. In fact, one of its handling quirks is a tendency to lift off prematurely, which must be counteracted by the pilot during the takeoff roll. Until the Dyke is moving at 70 to 80 mph and its gear is retracted, it is not ready to establish a solid climb. In fact, it is not until the best climb speed of 110 to 125 mph is reached that the airplane really comes into its own. At that point, owners report a distinct—and exhilarating—sensation of "getting up on the step" and a definite acceleration.

To an aircraft designer, the Dyke Delta represents a particularly clear case of why conventional airplanes are the way they are. Its claim of being a 180-mph four-seater is equivocal (in the terms of Wichita and Lock Haven its cabin would be described not as "four-seat" but as "single-seat with ample baggage capacity"), and it certainly could not be marketed against factory aircraft. Its performance is fine in some respects, substandard in others. Some of its curious characteristics—like the fact that it loses 400 to 500 fpm in rate of climb when the gear is extended or that the bottom 30 mph of its total flying speed range is only marginally usable—would disqualify it as a general market airplane, because general market airplanes will be flown by relatively unskilled pilots, and must above all else be safe and forgiving. Dyke himself wrote about the airplane, "The design theory of the Dyke Delta JD-2 is ultimate design efficiency [meaning] compactness of design, towability (eventually self-propelled roadability, I hope) and, of

course, high speed. I suffer with the JD-2 at low speeds, but gain bonuses at high speeds due to the whole aircraft lifting as an airfoil. The induced drag is no problem above 90 mph..." As those words seem to confess, the design is only ideal by certain personally assigned criteria.

For most homebuilders, however, comparative performance figures don't begin to tell the story of their love of their airplanes. Not every wife who is loved by her husband is beautiful, intelligent, or rich; nor is every husband loved by his wife handsome, witty, or brave. Dyke may or may not ever make the JD-2 roadable—it's unlikely—and compactness may or may not be a virtue in a rented T-hangar that was sized to accommodate a Bonanza. The criteria are fitted to the design. What pleases the pilot of a Dyke Delta is the

unique individuality that the airplane gives him. When he flies by, he turns heads. People talk about him. When he lands, a crowd gathers. If he makes three fast, flat approaches and goes around twice before getting the airplane onto the ground, it will be seen not as a flaw in the airplane, but as proof that he, the pilot, is skillful enough to handle a hot ship. He will get tired of answering the same questions again and again, but he will be crestfallen if people fail to ask them. That unique wing shape silhouetted against the clouds, an elegantly proportioned and symmetrical cluster of triangles and trapezoids strangely suggestive of a living thing, manta or moth, is a symbol of personal pride lifting itself above the "general market." Flashing past in the sky, the Dyke Delta is a complete, nonverbal explanation of what makes people build their own planes.

8

SONERAI & RV-3

Opposite
RV-3
Empty weight: 695 lbs
Wingspan: 19.9 ft
Length: 19 ft
Seats: 1
Cruising Speed: 185 mph
Range: 600 mi
Landing Speed: 50 mph
Engine(s): Lycoming
125 hp

Overleaf
Sonerai
Empty weight: 520 lbs
Wingspan: 18.7 ft
Length: 18.7 ft
Seats: 2
Cruising Speed: 140 mph
Range: 250 mi
Landing Speed: 45 mph
Engine(s): VW 1700cc

"Formula One" racing was invented after World War II to provide a class in which ordinary people could participate. It required a conventional tailwheel-gear airplane with a Continental 0-200 engine—at the time one of the cheapest engines available—and, for safety's sake, a fairly generous wing area. The rules were written in such a way that all Formula One airplanes had to come out looking pretty much alike. But as years passed the competition became more and more intense, and the refinements of streamlining, engine tune, and propeller design became so elaborate that a beginner stood no chance against the seasoned racers. Furthermore the Continental 0-200 gradually became a rather expensive engine, particularly when blueprinted, balanced, and brought up to race tune.

A new, more basic, more primitive class was needed, so Formula Vee, named in emulation of a similar class for economical auto racing, was instituted. Steve Wittman, famous as a race pilot from the thirties and the designer of many airplanes, was the first to build one under the rules of this new class. Formula Vee called for a slightly larger wing than that used in Formula One (75 sq ft versus 66) and a Volkswagen engine. Implicit in the formula was extreme simplicity of design and construction, and low cost. The key to low cost was the VW engine (which was much cheaper in 1968, when the class was launched, than today, but is still much cheaper than an aircraft

engine). Generally accepted—with certain reservations—as a suitable powerplant for an airplane, this engine had been in use on home-built airplanes of various sorts for some time, and had even been used on certificated airplanes in Europe. The VW engine, with its flat-opposed, air-cooled design, is quite similar to conventional aircraft engines. However, it lacks certain features customary on airplane engines—most notably, propeller flange, thrust bearings, and magnetos—and has to be modified to be used in a plane.

The most popular design in the Formula Vee class is the Sonerai, created by John Monnett. Monnett was a twenty-six-year-old high school art teacher when he succumbed to the compulsion to design an airplane (he had already built one) after attending the EAA Fly-In at Oshkosh in 1970. He wanted to build a Formula Vee racer and, what was more difficult, he wanted to build it in time to fly it to Oshkosh the following year. Building an airplane in a year is not easy, especially when you are also designing it. Design, particularly the kind that is roughed out at the drafting table and crystallized at the work bench, is time-consuming. You make a lot of false starts, and often have to assemble, backtrack, disassemble, modify, and reassemble. And though working against a deadline is exciting, it is also a torment, especially as the deadline draws near. The humor of your mistakes begins to escape you, and the recalcitrance and sadism of

inanimate materials sends you into a rage. You can't put down your tools and take a day or a week off, as you could if there were no deadline.

Monnett's design was extremely conventional and simple, resembling the Cassutt, a classic—but never very successful—Formula One racer. He permitted himself a few extravagances—the shape of the tail surfaces is a salute to the British Spitfire fighter of World War II—but on the whole clung to time-honored formulas: a rectangular, pop-riveted metal wing, a welded steel tube fuselage with fabric cover, a fiberglass cowling. In the interest of keeping it simple, Monnett called for only two sizes of tubing for the fuselage, and only .025 inch-thick aluminum sheet for the wings. He added one feature that suggests that he had more than pure racing in mind: folding wings.

The Sonerai didn't fly *to* Oshkosh in 1971, but it did fly at Oshkosh. One trip around the fly-by pattern, Monnett later said, was worth it all.

The Sonerai made a reputation as an airplane that could be built in a very short time—under 800 man-hours—and one that combined gentle flying characteristics with very economical operation. It also launched John Monnett's career as a designer for homebuilders. He published designs for a stretched two-seat version of the Sonerai, for a VW conversion kit for which he also supplied the parts, and, most recently, for a sailplane he calls the Monerai.

As it turns out, Sonerais have been built by more people than ever wanted to race airplanes. Most build them because of the look, the simplicity, and the low cost of both construction and operation, and because they are suitable for a certain kind of day VFR flying. The Sonerai is not a hot performer when compared to the Midget Mustang, which could be thought of as the airplane it replaces. It flies more like a Piper Cub, being extremely light (700 lbs gross weight) and having a large wing. It has a sturdy landing gear and can operate on grass strips of the kind that abound in the Midwest. So what most people do with their Sonerais is hop around between airports twenty miles apart on weekend afternoons, chatting with people here and there, and accepting the admiration and curiosity of chance encounters. All this is happening at airports that most people never know exist: semi-private farmers' fields, little town strips where crop-dusters work, patches of sod and turf far from cities and highways where never is heard a transmitted word, and navigation is by road and cow pond. And at the

end of the day, 6 or 7 gallons of fuel later, the wings fold neatly against the flanks of the fuselage, the tailwheel hooks to a bumper rig on the car, and the airplane trailers home on its own main wheels to be stored on one side of the garage, or in the barn.

In this context, the Sonerai fits neatly into the grassroots, cornball side of the homebuilding movement. It is a modest, unpretentious design for modest and unpretentious people who are not tormented by grand dreams. It is adapted to a kind of flying —"fun flying"—that is threatened by accumulating layers of government restrictions. With its auto engine—and incidentally, the ability to use auto fuel—it is in a category one step above the ultralights. It is also an example of a design for which the VW engine is well suited. Slow-landing, flat-gliding, it is a safe airplane in case of an engine failure, and the cautions against using VW engines in VariEzes, for instance, have less application to an airplane like the Sonerai which has a chance of faring better in a forced landing.

The Sonerai is stressed for aerobatics: the single-seater is tiny, less than 17 feet in both length and span, and sits low to the ground; the two-seater (which has about the smallest tandem cockpit in which two people could possibly sit) is heftier all around, and sits higher up. The single-seater can cruise at 150 mph on its 1600 cc VW, and lands at 50 mph. Both Sonerais are homely airplanes, but their partisans find in them the very image of speed and rakish beauty.

Although both airplanes use welded steel tube construction, they are within the reach of builders who don't know how to weld; they simply have to go out and learn how! It's not so difficult as it sounds; a high school extension course taken in the evening, or a few lessons from a friend are sufficient to start one gas welding, and from then on, practice quickly makes sufficiently perfect. The Sonerai's wing is extremely simple, and Monnett sells a fiberglass fuselage shell to replace the more complicated fabric cover. Various Sonerai builders, Monnett included, have reported all sorts of initial problems getting their VW engines to work properly, and the supersimple Lake (a.k.a. Posa) carburetor that Monnett recommends makes for hard starting. However, once the initial

problems are ironed out the airplane is quite troublefree. It has not led to a renaissance of air racing—it's doubtful that anything ever will—but it has contributed a lot to the perpetuation of grass-roots flying, which might be what Monnett, whether he realized it or not, actually had in mind.

Because the Sonerai is fast and inexpensive to operate, most people would consider it a highly efficient airplane. But "efficiency" —traditionally a key concept in airplane design—means different things to different people. When John Dyke defines the efficiency of his Delta in terms of compactness and roadability, for instance, he is enlarging the concept to include what normally is considered "convenience."

But definitions have to start somewhere; the mere fact that someone chooses to travel by

airplane, rather than by walking or riding a bicycle, suggests a definition of efficiency that is not absolute. Normal understanding of airplane efficiency includes good gas mileage and load-carrying ability, combined with safe flying characteristics. By this definition Richard Van Grunsven's RV-3 has always had the reputation of being a highly efficient airplane. In 1973 it won the Pazmany Efficiency Contest at the EAA Oshkosh Fly-In with an all-time high score. Thanks to that reputation, good reviews (especially one by Budd Davisson in *Air Progress*), and a sleek appearance, the RV-3 has come to be considered one of the best-designed, slickest homebuilts around. It is often cited as an example of how a small engine can achieve almost supernatural performance.

Actually, the RV-3 is a perfectly conventional airplane with a medium-size engine and performance that is good, but not shocking. It is relatively large for a single-seater, with a broad 90 sq ft wing and an empty weight of about 700 pounds. By comparison, Formula One racers, also single-seaters, have 66 sq ft wings and weigh at least 500 pounds empty. The RV-3, generously proportioned throughout, provides a more comfortable cockpit than most single-seaters. More important, it has a very low landing speed—48 mph—thanks mostly to its large wing area and landing flaps.

The airplane is practically a textbook example of conventional sheet metal construction. In fact,

with its rectangular wings, bubble canopy, and raked landing gear, the RV-3 is reminiscent of the T-18, that grandfather of simple metal homebuilts.

In designing the RV-3 Van Grunsven compromised between ease of construction and aerodynamic efficiency. Being a single-seater and therefore narrow, the airplane avoids the problem of severe fuselage convergence behind the cabin which led, in the T-18, to the polygonal turtledeck to which some people object on aesthetic grounds. The RV-3's comparatively high cruising speeds are remarkable not so much in absolute terms (a Vari-Eze attains comparable speeds with a smaller engine, and seats two people) as in terms of its very low stalling speed. As Dennis Polen did in his Special, Van Grunsven kept low speed behav-

ior in sight while designing for high speed.

The RV-3 is also, because of its large wing area, an airplane that profits much from careful attention to surface finish. Performance figures for different copies of the airplane are puzzling, however, as is often the case with homebuilts. The prototype with its 125-hp Lycoming engine has a maximum level speed of 195 mph, while Canadian Bill Pomeroy's 160-hp ship, which has won several awards for workmanship, attains just 198 mph. The 160-hp airplane climbs at 2,500 fpm, the 125 at 1,900. These differences seem quite small for a horsepower increase of nearly 30 per cent, and while no one suggests that Van Grunsven exaggerates his numbers or Pomeroy minimizes his, they point to the striking effect of different propellers,

different builders, and different flight testing techniques.

Overall, the RV-3 is an ideal single-seat design for homebuilders, combining the advantages of a proven engine and excellent low-speed capability, both of which contribute to safety, with beauty and high cross-country speed. The airplane is practical and enjoyable to use. In the hands of a competent pilot it can operate from the shortest grass or sod fields. Its only shortcoming, really, is the lack of a second seat.

Compared with the Sonerai, the RV-3 seems more like a "mid-sized car," and the Sonerai a "compact." The Sonerai is a relatively crude airplane, and significantly slower than the RV-3; but it can be built more quickly and more cheaply, and it can certainly be operated more cheaply. The RV-3's Lycoming engine is more expensive but more reliable than the VW engine; it also consumes more fuel and has much higher periodic maintenance costs. The RV is a cross-country machine, the Sonerai more of a puddle jumper—though the short-field capability of the RV and the good cruising speed and economy of the Sonerai allow each to share the turf of the other. The RV is certainly the more elegant airplane; and the person who has built one is ready to tackle any larger metal design. But he may not want to; the RV-3's excellent reputation and good looks inspire so much admiration and envy that the builder may find it hard to want anything else.

9

OSPREY II

Empty weight: 970 lbs
Wingspan: 26 ft
Length: 21 ft
Seats: 2
Cruising Speed: 130 mph
Range: 300 mi
Landing Speed: 63 mph
Engine(s): Lycoming 150 hp

During the thirties a lot of big airplanes were designed to land on water only. These "flying boats" were competing with ocean liners in international passenger transport. Because there weren't many big airports at that time, landing in harbors was more practical, and the ability to land on water added to the safety of ocean crossings. World War II changed all that. Airports multiplied, engines became more reliable, and large aircraft were able to attain such high speeds that the flying boats, encumbered by the drag of their floats and hulls, could not compete economically. They became extinct—except in a few places.

The unique thing about flying boats and amphibians (amphibians are flying boats that are also equipped with a landing gear) is that their value is so dependent on geography; in some places they're almost useless, in others they are indispensable. They come into their own in the re-mote areas of Canada and Alaska, where runways are few but lakes and rivers are many. Around Ketchikan in the Alaskan panhandle you don't see many landplanes and those that you do see are on their way to someplace else. The floatplanes (many of them not amphibious) that operate there are the only practical connection between the salt water highway that joins the islands and the coast, and the most practical mode of transportation to the inland lakes where sportsmen, trappers, loggers and prospectors come and go.

Obviously, the same criteria can't be used to judge seaplanes as land planes. The *raison d'etre* of the seaplane is simply its ability to operate from water; nobody worries about the aerodynamic drag of float installations, or about the ropes that are left dangling from wingtips and struts to assist in docking. Speed hardly matters, when the mere ability to go in a straight line from one place to another faster than a donkey can walk is already a huge asset. What does matter is water handling characteristics, corrosion resistance, durability, and short takeoff and landing ability.

Sport seaplanes, on the other hand, are designed with more than just practicality in mind; beauty and speed are also considered. These seaplanes are usually of the hull-in-the-water type, and generally have their engine mounted on a pylon high above the fuselage to keep the prop out of water spray. They are relatively difficult to dock, but the kind of use to which they are put most often involves beaching, which they can do easily, or taxiing up a boat ramp, or mooring to a buoy. They are used for getting to vacation spots on lakes and rivers, for fishing and camping, and for that amorphous but familiar category of recreation called "messing about in boats."

One of the most popular sport seaplanes is the Osprey II, designed by George Pereira, a Sacramento businessman. The Osprey II is an improved version of the Osprey I, a single-seat flying boat. It was equipped with folding wings, because the only practical way to store it was to trailer it home, and it had no landing gear, being a flying boat, not an amphibian. Pereira found this lack of landing gear to be very impractical; it is often difficult to get aviation fuel on bodies of water, and one ends up longing for the ability to land at airports. He ended up selling the Osprey I

The canopy, opening like the upper jaw of a crocodile, permits the pilot to climb out of the cockpit and down over the nose to fend off a buoy or hop onto a dock.

to the U.S. Navy—they considered using it as a river patrol plane during the Vietnam War—and building a second seaplane which was equipped with retractable landing gear and two seats. He spent 1,300 hours and about $4,200 building the prototype, which he completed in 1974.

The Osprey's configuration is a classic one for small amphibians. Although the height of the pylon-mounted four-cylinder Lycoming engine above the fuselage seems exaggerated, there is a certain minimum practical diameter for a propeller, and that diameter determines the height of the pylon. If the entire airplane is small, then the height of the pylon seems excessive.

The Osprey is constructed primarily of wood, plastic foam, and fiberglass, somewhat in the style of a model airplane. The drawings of the fuselage sides are laid out on a long work table and used as templates for gluing the frames; the trusses are then stood on edge and the cross-members glued into place. The wood used throughout is strong Douglas fir, available from lumber yards, rather than less easily obtainable aircraft spruce.

The method of shaping the hull is interesting. The cabin has a flat floor, with a keel structure protruding beneath its centerline. After the fuselage structure is assembled, it is trailered down to a home insulating company where insulating foam is sprayed on the underside of the cabin floor, bury-

Part of the amphibian experience is the pure pleasure of being on the water.

ing the keel. Back in the shop, the mass of foam is carved to the proper full contour, which resembles that of the bottom of a boat, and is covered with several layers of fiberglass. Though light in weight, this type of structure provides a stiff, durable hull which is leakproof even if its fiberglass skin is punctured, and a flat cabin floor without bilges to collect water and dirt. The structure of the compound-curved cabin roof is similarly ingenious. Polyurethane foam is expanded in place over a supporting layer of paper, carved to shape, glassed, and then turned over and shaped and glassed on the inside. This kind of construction, by the way, is typical of the unconventional but highly practical methods homebuilders have developed for applying composites and plastics to airplane structures.

The Osprey II also has a big, flatwrap plexiglass windscreen which hinges at the top edge, opening like a crocodile's upper jaw to permit the pilot to climb down the nose of the airplane onto a dock or beach. The very compact, retractable tricycle landing gear is operated like that of the early Mooneys, by a handle between the seats. Engine controls are on a console between the seats, not, as is customary in amphibians with pylon-mounted engines, on the ceiling.

The wings and tail surfaces are of fabric-covered wood construction, but rib-stitching is omitted, thus simplifying the work. The outboard floats, in another expedient application of foam-and-fiberglass construction, consist of a thin plywood profile on both sides of which slabs of foam are glued, shaped and fiberglassed over.

When completed, the Osprey is an attractive amphibian which, according to reports, turns in a very satisfactory performance. It has a cruising speed of about 130 mph, a stall of 55 mph, an 800 fpm rate of climb at gross weight, and good water handling characteristics.

Though the Osprey has the advantage of wood and composites construction which resists water damage well, potential builders should be aware of the relatively high maintenance burden of metal parts, such as landing gears and wheel bearings, which are exposed to water. The landing gear is most often exposed to water when the airplane is being run up a boat ramp, or when, during a stiff breeze, it is lowered into the water as a sea anchor. Salt water is particularly hard on parts, and if the wheels get into salt water, the bearings, even if they are of special stainless steel, must be repacked after each exposure.

Another problem—not so much for an amphibian like the Osprey as for a pure seaplane—is that not all bodies of water are available for seaplane operations. Many lakes, even though they permit boats and water skiers, ban aircraft. Consequently, there are large sections of the country where, even though there are lakes and rivers, there is no place to land on water. On the other hand, there are places like the Puget Sound area and the Alaskan panhandle where only amphibians and seaplanes can be used; but for a person living in Arizona or Missouri, they are cold comfort. Proximity to the ocean is not necessarily a help either; sheltered harbors often prohibit airplanes, and the open sea is too rough for a small seaplane like the Osprey.

These considerations are important because the ability to operate satisfactorily on water can be bought only at the price of some performance in the air. The same engine that drives the clean and compact Osprey at 130 mph would propel a similarly clean and compact two-seat land plane at nearly 200 mph. If an amphibian is going to be used on water very rarely, it's worth wondering whether this type of plane—despite its versatility—is really the best choice.

But in defense of the seaplane it must be said that there is something nearly magical in flying to and from water. It is a sensation that one cannot imagine without having experienced it. Rather than giving up the freedom of the air when you land, you find yourself trading one freedom for another. It is worth many hours of slow cruising to be able occasionally to come to rest on water, shut down the engine, lift the canopy, and find oneself inches from the blue-green surface, rocked by gentle swells and listening in the cool stillness to the restful slap and giggle on the hull.

Water handling is an important part of the seaplane's character. The takeoff run begins slowly; then the plane raises itself, like a speedboat, onto the "step." Hydroplaning, it accelerates rapidly and lifts into the air, trailing streams of vapor.

10 CHRISTEN EAGLE II

Empty weight: 1025 lbs
Wingspan: 19 ft 11 in
Seats: 2
Cruising Speed: 165 mph
Range: 380 mi
Engine(s): Lycoming
AEIO-360-AID, 200 hp

Frank Christensen likes to go first class. So do his customers, a list of whose names, as some have remarked, reads like the Fortune 500. Frank's company represents the lovable side of capitalism: a fine product beautifully presented; happy and affluent customers whose needs are serviced with extravagant care, as in a fine restaurant; the personal touch; and few worries.

Christensen made his own fortune manufacturing equipment which in turn was used to manufacture semiconductors; he sold his business when he was 32 and retired with—if his subsequent activities are any indication—a rather tidy nest egg.

Frank's obsession is aerobatics, that branch of aviation in which grace is bred with geometry to produce surprising and exciting

flight. Aerobatics is also a competitive sport. It supports an annual world championship that has taken place in cosmopolitan venues both capitalist and communist (it is state supported in Soviet-bloc countries), and it has brought into being a unique type of airplane answering to special requirements. An aerobatic airplane must have crisp, effective controls and high maneuverability; a rapid roll rate; vertical penetration (meaning the ability to fly straight upward while maneuvering without rapidly losing speed); the ability to fly well while inverted; and carefully designed stall and spin characteristics.

One of the most successful aerobatic airplanes is the Pitts Special, a small biplane that actually started out as a homebuilt but eventually found its way into limited production. With its small

size and powerful controls, the Pitts has the necessary agility; with its big engine—normally a 180- or 200-hp Lycoming, sometimes fitted with a constant-speed propeller—it has good vertical capability; its deep, short fuselage is good for "knife-edge" flight—straight, level flight on its side. European connoisseurs complained (partly, no doubt, because it was not invented there) that the Pitts was too small and quick, and was therefore difficult to judge in competition and perhaps a bit too easy to look good in. The U.S. team won a world championship in Pitts Specials in France in 1972, though judges complained that the airplanes lacked grace and style, and more resembled circus acrobats than Olympic gymnasts.

Frank Christensen recognized that even the Pitts had some shortcomings, as all airplanes do. He wanted to redesign the Pitts and offer it as a kit, but Curtis Pitts resisted. Undaunted, Christensen designed his own plane instead.

The Christen Eagle is the result. It is in essence a glorified Pitts Special—no one denies that—but it has significant superiorities to the Pitts. It is also the ultimate model plane for grown-up kids.

All the parts come in tidy crates, each individual part or group of parts bagged or sealed in plastic against a cardboard backing. Everything is numbered. Attached to the inside of the crate lid is a razor blade to help cut

An airplane like the Eagle is supposed to fly as well upside down as right side up. This causes the engine no problems; Christen Industries had made a reputation with its inverted fuel and oil systems long before the Eagle was designed.

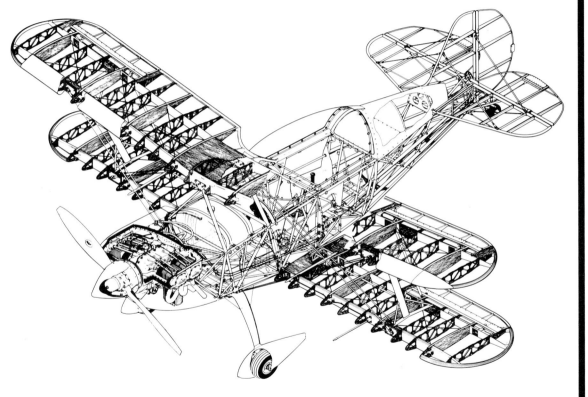

The construction is absolutely classical—which is also to say, very complicated. Volumes of step-by-step plans make it relatively simple, but not quick.

open the plastic. The instructions (which are set up and computer stored in an electronic text editor for ease of modification) go step by step through the building process, giving expertly drawn exploded views of each small assembly. These pages alone fill 32 thick volumes in hard ring binders. All welded assemblies are done at the factory. Christensen's idea was that the well-off professionals who would be building his kit wouldn't have time for a lot of scrounging and legwork, ordinarily two of the great time consumers of home-built projects. Once they started the Eagle, he said, they would never have to leave the garage.

The price of Christensen's perfectionism is high: with an engine, the kit runs to approximately $27,500. Furthermore, though it is easy to build, it takes a lot of time. Estimates of assembly time run from 1,200 to 2,000 hours.

A lot of famous pilots flew the prototype Eagle (with its astonishing nine-tone paint job, the only part that could not be supplied with the kit) and agreed that yes, it was better than the Pitts. It was more comfortable and roomy. It had that long bubble canopy. It was cleaner and faster and had better vertical penetration. It was easier to land. It did superior snap maneuvers. The instrument panel was easier to see because it was farther from the pilot (whose eyes are normally focused on the horizon). The airplane was better balanced, though at the cost of moving the

pilot farther behind the wing, from where it is more difficult to use the wing trailing edge for position reference. And its simple spring-steel landing gear was easier to maintain than the Pitts' bungees, which have to be replaced periodically.

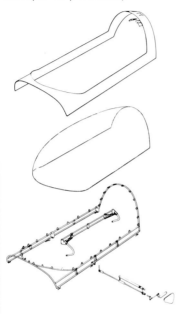

But where was Frank Christensen going to find his customers? It is difficult to imagine wealthy professionals—doctors, lawyers, airline pilots and captains of industry—laboring for years over a vast and complicated model airplane. And they hardly seem likely to indulge in free-for-all aerobatics, let alone serious competition. Did Christensen really think that there were enough people like himself out there to make a market? He did,

and he apparently was right. Hundreds of orders for Christen Eagles have been placed, and Christensen has found himself in the position of being unable to deliver kits rapidly enough, of having to ask his customers to please bide their time.

Christensen also developed the Eagle I, a 260-hp single-seat version of his airplane for all-out world-class aerobatic competition; three of these have been delivered to the Red Devil aerobatic team to replace their Pitts Specials. And he has plans for a whole line of kit-built airplanes of different types. On top of that he wants to produce a book tentatively called *Christen's Aerobatics*, which, like everything else he does, should be the last word on the subject.

It has been suggested that all those people bought Eagles out of regard for Frank; but although he's a nice fellow, he could hardly have that many close friends. No, the fact is that his aim was good, unlikely as its direction seemed. Building airplanes is obviously not just a hobby for handymen with an abundance of spare time, or a distraction for the retired. Anyone might be willing to take a crack at it, if the tone of the project is right. Frank found the right tone for the Rolls Royce set: an airplane kit for the perfectionist, immensely expensive but lovingly designed and impeccably set out. With the Christen Eagle, an airplane in the Gucci and Hermes tradition, homebuilding finally acquired its own Upper Class.

69

11 | ROTORWINGS

Opposite
Bensen Gyrocopter
Empty weight: 247 lbs
Rotor diameter: 21.7 ft
Seats: 1
Cruising Speed: 60 mph
Range: 100 mi
Engine(s): McCulloch
72–90 hp

Overleaf
Scorpion Two
Empty weight: 790 lbs
Rotor diameter: 25 ft
Length: 20.5 ft
Seats: 2
Cruising Speed: 75 mph
Range: 150 mi
Engine(s): Rotorway
RW 133, 120 hp

Crowds gather to watch
a Scorpion owner set up
his ship. Note the air-
brush work on the nose;
a Scorpion in flight, the
not-quite-impossible dream.

Lil' Susie

There are not many two-seat
gyrokites around, but they
represent an excellent way
to obtain dual gyrocopter
flight instruction before
taking your life into your own
untrained hands.

No doubt the crowning aeronautical fantasy is the helicopter.
It is the airplane freed of its last
constraint, the need for runways.
With a helicopter you can go
from anywhere to anywhere, or
at least so the fantasy goes. From
the roof of your house to your
downtown office; from there to
your mountain cabin; to the
marina; to Aunt Bea's farm; to
the lake for some fishing; to the
other side of town for a movie.
With a helicopter, you suppose,
life will be different. Angels
couldn't have it any better.

Manufactured helicopters are expensive—very expensive. They
are a collection of highly-stressed
moving parts, each part manufactured with precision to close tol-

erances and carefully inspected,
then periodically reinspected or
replaced. Helicopters abound in
opportunities for metal fatigue.
They must remain in delicate ad-
justment and balance. If a part in
the rotor system fails, it's all over.
The cheapest factory-made helicopter costs $40,000.

But could you build one?

Indeed you could. You could build
either a helicopter or, much more

simply and cheaply, an autogyro.

The theory behind rotary-wing aircraft is a little difficult to understand. So if you don't understand it, take it on faith that a helicopter's rotor won't stop turning if the engine quits. Just as an airplane continues gliding forward and downward after a power loss, the helicopter rotor keeps gliding around in circles as the ship settles downward, levels out just above the ground, and makes a soft landing.

Once you accept that a helicopter rotor can glide without an engine driving it, you know how autogyros work. An autogyro is a helicopter with a freewheeling rotor. The whole contrivance is propelled forward by an engine driving a conventional propeller, like an airplane's. The beauty of the autogyro is that it does away with the complicated rotor controls, transmission, and clutch used in helicopters. Its rotor is a single rigid assembly, free to teeter at the top of the mast. By

tilting the entire rotor assembly, the autogyro pilot causes the ship to turn, speed up, or slow down. He gets lift as an airplane pilot does, by increasing power and holding a steady speed, not, as with a helicopter, by increasing the pitch of the rotor blades.

The autogyro has one significant advantage over an airplane. When an airplane slows down below a certain speed, it can no longer fly. An autogyro can be standing still, but so long as its rotor blades are still spinning at high speed, it can produce lift. Lift is independent of forward motion—to a certain extent. In a helicopter it is completely independent. An autogyro needs forward motion to start the rotation of the rotor, but momentum will keep it spinning after the autogyro has slowed down or completely stopped moving forward. So an autogyro can land standing still; and if it has a little motor to accelerate its rotor before takeoff, it can take off in a very short distance, though not vertically like a helicopter.

The Bensen Gyrocopter, an autogyro of extremely simple design— a kind of flying lawn chair—was for a long time the most popular type of homebuilt aircraft. Bensens outnumbered all other types of homebuilts put together. Very cheap—$1,500 was a common price in the sixties—they used a two-stroke, four-cylinder McCulloch target drone engine (light, cheap, and fiercely powerful) to drive an airframe that consisted of little more than a few pieces

The gyrocopter rotor (lower left) is simple: a few aluminum angles and plates bolted together. The rotor of a true helicopter (above)—as simple as the Scorpion's may be—is vastly more complex, with precision fits, bearings, and numerous carefully machined parts. The payoff: the helicopter can take off vertically, the gyrocopter cannot.

of square aluminum tubing bolted together. Besides the engine installation, which was pretty basic, the only demanding part of the design was the rotor, and that was normally bought ready-made from Bensen suppliers.

Igor Bensen himself became a minister of the Gospel after many years of promoting his contraption, some said so that he could give last rites along with each Gyrocopter. This macabre witticism referred to the little craft's safety record. Accidents were frequent, ranging from pilots stepping out of their machines after a landing and being killed by the still-spinning rotor to structural failures, engine failures, and loss of control in flight followed by separation of the rotor blades. This last scenario, a common one, was called a "stall" by FAA accident reports. But the Gyrocopter, like a helicopter, is characteristically incapable of stalling, and the real explanation was

73

probably that the pilot pushed the stick forward too abruptly, putting a negative G-load on the rotor which caused it to bend down and strike the rudder. At any rate, Gyrocopters got a reputation for being dangerous, which was not helped by the fact that they were almost completely unregulated. The only requirement for flying a Gyrocopter is a student permit, which is no more than a medical certificate with an instructor's endorsement. No dual instruction is mandated, so a lot of people who have never flown anything else before have tried their luck in Gyrocopters.

It's too bad, because flying one is fantastic fun. With as much power per pound as a race car, it accelerates and climbs rapidly, responds instantly to the throttle, and maneuvers beautifully. Like airborne dirt bikes, Gyrocopters leap over obstacles, soar, pirouette, parachute, plunge, catch themselves and finally flutter

down, like big birds, to gentle landings. Out in the desert and the countryside they play in the air with the grace and freedom of swallows. There is simply nothing else like them.

But the dream of point-to-point transportation will have to wait for something else. Gyrocopters can't go very fast, and their motors vibrate in an unpleasant way. A California dealer named Ken Brock once flew a Bensen across the country, but he is certainly the only person who has ever done so; and he chose a different way to get back.

A much more sophisticated, elegant and costly, but generally more versatile and satisfactory approach to vertical flight is the helicopter. There are many obstacles to building a homemade helicopter, but one manufacturer has made a success of selling kits, engines, and flying lessons all in one huge package.

The manufacturer is Rotorway, located near Phoenix. The guiding genius behind the enterprise is an enthusiastic, persuasive man named B.J. Schramm.

Schramm's present offering, the Scorpion Two, is a genuine helicopter. He sells it for $13,500 in the form of a kit which takes most people a year or two to assemble. The elegantly sculptured steel-tube frame comes tack-welded together; the buyer must complete the welding, then install the Rotorway 133-hp engine (which is derived from a water-cooled racing version of the Volkswagen engine), the reduction system (which consists of chains and V-belts), the rotor blades (which come ready-made) and the cabin and controls. Good mechanical skills are necessary, and so are medium-level shop tools like a drill press. On the other hand, there is more assembly than fabrication, and the plans combine exploded views with a tab-A-in-slot-B narrative text. As kits go, the Scorpion is marginally close to violating the FAA rule that the builder of a kit airplane must do at least 51 percent of the work; but the FAA has approved it.

Once he has built his Scorpion, an instructor's endorsement in his logbook is all the owner needs, besides his medical certificate/student license, to fly his helicopter solo. To carry a passenger, however, he must go a step farther; he must have a private pilot's license and helicopter rating.

Rather than limit his market to holders of helicopter ratings (which are very expensive to obtain), Schramm started a school at his Phoenix facility to train his customers to fly their machines. The curriculum is intense. The student spends eight hours a day for a week in ground school and hover training. Then he goes home and practices hovering with his helicopter tethered to the ground. When he has logged twenty hours, he returns for the "climbout" portion of the course, in which the rest of helicopter flying is covered. Then he is expected to go home and practice again. He is still entitled to fly solo only.

To carry a passenger, he must pass an FAA helicopter flight test for a private license. Some FAA examiners will accept a flight demonstration, with them remaining on the ground and watching, in lieu of a check ride; others may be willing to go along in the experimental helicopter (though most probably would not). In a few cases no examiner this cooperative may be found, and then the owner has to pay for enough dual instruction in a certificated helicopter to get recommended for the flight test and pass it. All this, mind you, in order legally to carry a passenger.

Like gyroplanes, helicopters are terrific fun to fly. The freedom and rapidity of motion that they provide is exhilarating. But with helicopters as with autogyros, balloons—in fact every type of flying machine—there are legal obstacles to complete freedom of movement. It's against the law to land a helicopter anywhere except in designated airports or helipads. Many helipads, especially in city centers and atop buildings, are privately run, and because of their insurance requirements their owners will not let homemade helicopters use them. Out in the countryside you can't land on private property without the owner's permission. Nor can you even operate legally from your own driveway, if you have neighbors, without laying an elaborate groundwork of permissions and authorizations, because the helicopter is an obvious noise nuisance and, in the eyes of many, a hazard to people underneath it.

Brochures advertising the Scorpion and other helicopters feature pictures of couples camping under the stars beside their ships, or taking off from front lawns, or landing beside apartment houses. These pictures are somewhat misleading; reality is not so idyllic. Nevertheless, helicopters are pretty neat machines. People have a lot of fun with them, and even if they don't end up commuting in them, owners usually aren't disappointed. Some Scorpion builders are lost in enthusiasm for their projects even though they have never done more than hover them in the tethers.

Rotorcraft flying may not free you from the tyranny of the car, but like surfing, hang gliding, parachuting or backpacking, it can give your life some new high points. Just go on a sightseeing hop once a month in a Scorpion, or do a little hedgehopping in the foothills with a Bensen. You'll find some sublime moments; and you won't get bored.

Learning to fly the Scorpion involves hours of hovering hesitantly in place, usually with the skids tied to the ground in case things get out of control. Schramm insists on supervising the training of all his clients.

12

ULTRALIGHTS

Empty weight: 130 lbs
Wingspan: 34 ft
Length: 7 ft
Seats: 1
Cruising Speed: 55 mph
Range: 90 mi
Landing Speed: 25 mph
Engine(s): McCulloch
12 hp

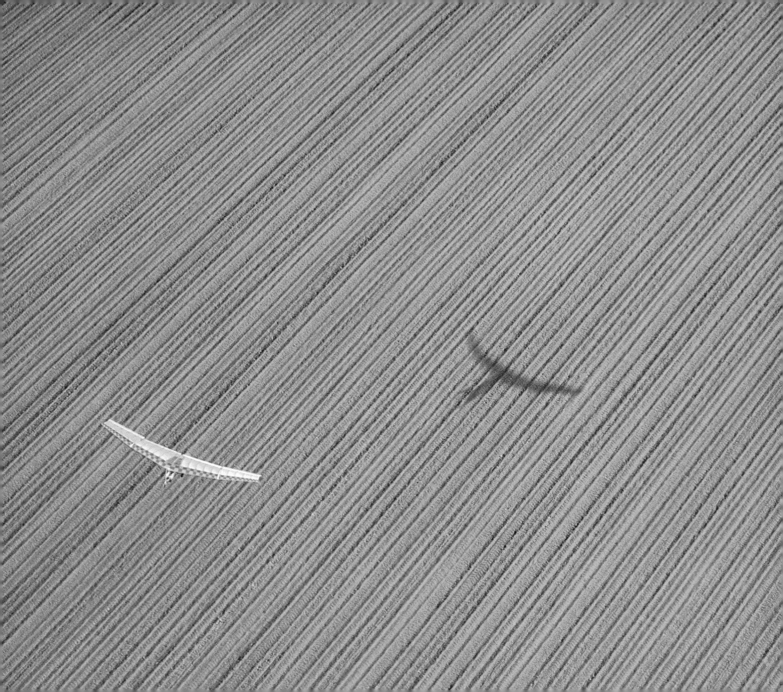

The most rapidly growing category of homebuilts is that of "ultralights." Ultralights gained entry into the kingdom of airplanes when flight enthusiasts began attaching various sorts of engines to what were essentially sophisticated hang gliders. Hang gliders themselves are not technically airplanes, according to the FAA definition of the word. The FAA decided several years ago not to involve itself with the whole area of hang gliding and excluded from the normal licensing requirements all aircraft which were, or could be, foot-launched. Anything which used the pilot's legs for landing gear in any sense—even if it had supplementary wheels for landing—was not an airplane, and could be flown without a license. This is a heartwarming example of governmental nonintervention, and shows that some bureaucrats

are motivated by something other than a mad desire to expand their spheres of influence.

Because hang gliders are very light and slow moving, they can be kept aloft by very small engines; and their engines need not be particularly reliable, since they are gliders to begin with and can get along quite nicely with no engine at all. Generally, any hang glider of relatively high aspect ratio—which leaves the triangular "Rogallo" type out of consideration—can be powered with one or two motors of 3 to 10 hp. All ultralights are similar, however, in having large wings of very light construction, often covered with Dacron or a thin, tough, transparent plastic film called "mylar"—the material of which recording tapes are made. The pilot usually hangs in a sling beneath the wing, and

the "cockpit" is arranged, like those tailgate-equipped pyjama bottoms, so the pilot's lower end can optionally dangle down below the aircraft. Some ultralights are tailless, while others have conventional or relatively conventional tail surfaces. The engines are little chain-saw engines, or similar miniature single-cylinder types, and are usually mounted on pylons at a short distance from the pilot.

Why do ultralights exist? Supposedly to put people in touch with the basic sensations of flight—the flight of birds, with the wind in one's face and the very movements of the body providing some or all control. The earliest Rogallo hang gliders, belated inheritors of the tradition of Lilienthal and the Wright Brothers, flew this way, with the weight and inertia of the pilot's own

Ultralights are versatile; they are airplanes, hang gliders, self-launching sailplanes. Prop stopped, a Mitchell Wing wheels slowly earthward.

79

body providing flight control. In the early seventies, however, designers like Vollmer Jensen, who understood the basic principles of flight efficiency, developed hang gliders with large high-aspect ratio rigid wings and conventional control surfaces. Rigid-wing hang gliders, despite their cumbersomeness, fragility, and difficulty of construction, were more than competitive with the much simpler, but potentially more dangerous, Rogallos. At the same time a few super-simple home-built airplanes, like Bob Hovey's little biplane, Whing-Ding, approached the same region from the other direction, putting the pilot into the breeze, reducing the size and complexity of the airframe to a minimum, and using a go-kart engine for propulsion.

Once the efficiency of hang gliders had reached a certain level, and the technology of superlight structures was well established (reaching its zenith in August 1977, after decades of

effort, in a controlled mile-long flight by a man-powered aircraft), ultralights suddenly burst on the scene. Ultralights seem to combine the advantages of motive power with those of bird flight. Of course, no bird ever made so much noise as a chain saw engine, so an unexpected feature, the raucous snarl of the dirt bike, has somehow intruded itself upon the ornithophile ideal.

The engine can be viewed, like that of the self-launching sail-planes, as a substitute for the tow plane; or like the engine of a powered hang glider, as a substitute for the hill from whose summit one leaps. But it is not hard to imagine—and experience bears one's suspicions out—that many afficionados of ultralights will spend more of their time chasing one another around at full throttle than climbing up, shutting off their engines, and floating down in silence. The noise and ferocity of the engine, like that of the dirt bike, will be more seductive to pilots weaned on the sirens and gunshots of television than will the meditative silence of powerless flight with its disturbing resemblance to a home power outage.

The latest development is that of electrically-powered ultralights possessing the virtues of silence as well as power. Their output is limited, but a few pounds of rechargeable batteries and a small, efficient motor permit a climb of several minutes followed by a long glide—a good hang glider can remain aloft indefinitely in updrafts and thermals—and, after landing, a recharge from a car's alternator.

Also imminent is the solar-powered airplane, an ultralight whose surfaces will be covered with cells converting sunlight into voltage to drive a motor keeping the aircraft aloft. The larger a wing, the less power is needed to keep it airborne. The relationship between the power and wing area required for manned flight and the power per square foot available from sunlight just happens to be such that large, very light airplanes of the order of the current ultralights, could be kept flying by light alone. With the eventual realization of that sublime development the word "ultralight" will have attained its most complete—and delightful—significance.

The Mitchell Wing is light but cumbersome. Using quick-release pins, its frame fuselage is secured to the wing, which in turn folds up for storage and transportation. It has to be big in order to fly with so small an engine.

13 BUILDING YOURS

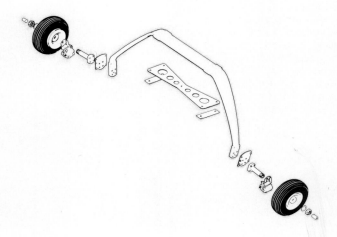

Deciding Which Plane

Because there are so many homebuilt kits and plans on the market—not to mention the possibility of designing one's own airplane from scratch—the decision of what to build, once you have made up your mind to build *something,* can be difficult. Sometimes a person is attracted not by the idea of building *some* airplane, but by one particular airplane that he has seen or flown. For that person the decision is easy; he wants that plane, and he starts working on it. But once you decide simply to have a go at homebuilding, selecting the specific project requires some soul-searching.

There are several important factors everyone should consider: purpose, ability, time and money. Purpose, the reason why you are building a plane in the first place, is the most difficult aspect to be realistic about. One treats oneself to all sorts of notions about how life will be with the airplane; in most cases, however, it is not likely to revolutionize one's existence. So it is well to keep one's ambitions moderate. It may be seductive to imagine yourself roaring around at 300 mph in your Brokaw Bullet, but if the Brokaw is going to take years to build, and its engine (to say nothing of the rest of it) will cost so much that you won't be able to support it once it's done, then it would be better to settle for a less megalomaniacal design.

Your ability as a craftsman is also important. If you have never built anything before, it would be unwise to start off with a project on which your life may eventually depend. Perhaps you should begin with something less demanding than an airplane. And if you have always been a slipshod workman, don't think you will mend your ways for this one all-important project; the temptation to cut corners, if one experiences it at all, wins out in the end. If, on the other hand, you have done a lot of building and are at home with tools and plans, then selecting the type of construction is less important than choosing a design. If you have worked a lot with glue and wood and are equipped to do so, you might prefer a wood design, but it's not difficult to switch to another medium if you should opt for a design made of metal or fiberglass.

Costs

You have to be prepared to see your wildest estimates of time and money exceeded by a factor of two or three. If the super-simple VariEze typically costs $10,000 to build, other projects will probably cost at least that much. The published costs for old designs don't mean much, since the prices of materials rise quite a bit from year to year, and even a builder who just finished his project may have paid for most of the components five years ago. For many of the popular new designs prefab parts are available—if there is not a complete kit—and builders often end up spending extra money to save themselves the trouble of building special parts, like canopies or land-

ing gears. In general, such money is well spent. A part like a plexiglass canopy, which involves a large investment of money and effort in tooling as well as a risk of producing a defective part on the first try, may actually be cheaper to buy than to build, even though the list price may seem high.

Types of Construction

In choosing the type of design, it's a good idea to think hard about the utility of the airplane. A few people who have "just for fun" single-seaters do use them quite a bit, going up on summer afternoons after work for some loops and rolls, or visiting a nearby airport on the weekend. Some people travel long distances in their sport airplanes. But most sport airplanes spend a great deal of time gathering dust. Airplanes are not much different from cars; if you had a car which was suitable for nothing but going out for a spin alone now and then, how often would you use it? Even those people who enjoy driving end up combining pleasure with utility. It is therefore advisable to pick a design of the widest utility; otherwise flying will end up competing with your other leisure-time activities—and those of your family—and probably losing.

Two-seaters stand a better chance of being useful than single-seaters; planes with luggage space are more useful than those with none; and so on. Remember that you are much more likely to have to shoehorn your airplane into your everyday life than you are to refashion your life around your airplane.

All-metal construction, with some nonstructural fiberglass parts, is now the standard of the commercial industry. It is a good type of construction for a homebuilt as well. Aluminum sheet is readily available, and the techniques of using it are easy to learn. The great advantage of this type of construction—and of fiberglass composites—is that once you have built the structure, you have built the airplane. The disadvantage of steel frame and fabric cover is that you have to build three airplanes: the skeleton-like framework, the wood and metal shape-giving structure over that, and finally the fabric cover over everything. What sounds like a simple step-by-step process is actually very time-consuming, involving innumerable steps and a variety of skills. Metal is simple; you form spars, ribs and skins, rivet them together, and paint.

The typical metal wing consists of a strong main beam called the "spar," airfoil-shaped ribs running fore and aft, a rear spar about three-quarters of the way back to the trailing edge, and skins. Ailerons and flaps are built as separate structures and hinged to the rear spar.

An all-metal fuselage takes
shape in an outdoor shop.
Simple in appearance, it
represents months of effort.

The principal fabricating operation involved is shaping the ribs, which must have flanges bent at a 90-degree angle to their surfaces along the perimeter. These flanges are later used to rivet the skins and spars to the ribs. The flanges are formed by sandwiching the sheet aluminum "blank," which is in the shape of the rib with an additional 5/8 inch of material around the edge, between two pieces of wood the shape of the final rib. One of the pieces of wood is a fine-grained hardwood (metal, masonite, or other hard, durable materials can also be used) with its perimeter smoothed and a radius sanded where the metal will be bent over it. The excess metal is then hammered down with a plastic hammer, a rawhide mallet or some other type of tool that is softer than the aluminum. If annealed aluminum, such as 2024-0 alloy, is used, there is a hammering technique for

shrinking the excess material out of the flange, eliminating wrinkles; if half-hard (6061-T4) or hard (2024-T3, 6061-T6) sheet is used, excess material is taken up by "fluting" the flange at intervals, that is, by making a neat, shaped scallop in the flange every couple of inches. Without this shrinking or fluting process, the rib will either have a wrinkled flange or will bow—both unacceptable conditions. The process sounds complicated when it is described in words, but with a little practice one can turn out a complete rib, from cutting out the blank to flanging the lightening holes (if any), in half an hour or so. Ribs are typically spaced at 12- to 18-inch intervals, and each is in three longitudinal sections (nose, between spars, and aileron or flap); making a complete set for an average wing should take 20 to 30 hours, the trailing edge ribs being much smaller and simpler

to form than the others.

Rectangular-winged homebuilts are ordinarily designed to take advantage of the standard 4-foot width of aluminum sheet. A piece of the proper length is wrapped around the leading edge (the leading edge radius can be bent by folding the aluminum sheet over, sandwiching it between two boards and kneeling on the boards), and several holes are drilled in the ribs and rear spar to hold the skin down tightly over the frame. Temporary assembly is done with sheet metal fasteners called "Clecos," and a simple wooden jig is usually used to hold everything in the proper shape, avoiding twists. Then the remaining holes are drilled—a thousand or more of them—and the wing is disassembled. Metal chips are cleaned away; the rivet holes are individually dimpled if the riveting will be flush; and the metal parts are cleaned and primed with anti-corrosion treatment. Now the wing is ready for riveting. An FAA inspector comes to check the workmanship before you rivet the wing shut.

Riveting is traditionally done with solid aluminum rivets and a rivet gun. The rivet gun is an air hammer which requires an air compressor. Used, a complete riveting rig will run from $200 to $500, including an assortment of "rivet sets"—the business end of the rivet gun—and smooth steel bucking bars against which the "shop head" of the rivet is formed. Because of the cost and the need to have a variety of sets and bucking bars, it's a good idea for several builders to share the same equipment.

A popular alternative to conventional riveting is so-called "blind" riveting, in which a hand-operated squeeze tool is used to insert and upset a specially made rivet. Blind riveting systems, of which many are available, have definite advantages: silence, suitability for one-person use (in some structures it is impossible for one person to hold both the rivet gun and the bucking bar) and an ability to fasten hard-to-get-at structures from the outside. Blind rivets are expensive compared with conventional solid rivets, but a comparable amount of expense is saved on the compressor and air hammer.

Finally, a few perfectionist builders go over the outside of their metal structures and smooth each rivet head with an epoxy filler which is then carefully feath-

ered into the surrounding surface. When the metal skin is as smooth as though it had no rivets in it at all, they prime and paint. Total time—200 to 600 hours for a complete wing.

Fiberglass composite construction, as typified and practically monopolized (though not for long, probably) by the airplanes of Burt Rutan, is the quickest and simplest method of all, though it may seem the most unfamiliar and discouraging. You do not need to fabricate, drill and rivet together dozens of separate ribs, frames, spar caps, spar webs, skins and hinges, as you do in metal. Instead, you make simple foam cores and cover them with epoxy-wetted glass cloth.

The cores are made in two ways. For wings and control surfaces the cores are cut out of blocks of light-blue styrene foam with a taut wire which is supported by two insulators about four feet apart and heated by an electrical

current. Using wooden templates at each end of the foam block, two people can learn to make very smooth and accurate cores with a little practice. For fuselages, cowlings and wingtips—known as "compound curved surfaces"—urethane foam is used; it is roughed out with a butcher knife or hacksaw blade, and smoothed with sandpaper.

When the core is ready, fiberglass cloth is cut to the appropriate sizes. While one person prepares batch after batch of carefully proportioned epoxy, the other lays the successive laminations of cloth down on the core and wets them out with epoxy. Using brushes and squeegees, both workers then spread out the epoxy, wetting the cloth thoroughly and removing all excess, which adds weight without adding strength.

While metal or wood work proceeds by gradual stages, foam-and-glass construction requires some advance preparation and then a number of intense hours-long sieges of laminating in which several people work concertedly, and which produce, at the end, finished wings or fuselage panels.

The surface finish of foam-and-glass parts is always rough. Since aerodynamic efficiency requires smooth surfaces, you subsequently spend quite a bit of time with lightweight fillers and sandpaper, smoothing and contouring the surface. The amount of work and the weight of filler required are in inverse proportion to the care with which the cores were

The BD-5 wing, here ready for closure, is typical of the metal idiom, though it differs in some respects: for instance, the use of a tubular spar, and the use of a sealant which is applied in this case because the wing also serves as a fuel tank.

made and the lamination done in the first place. Total time for a glass wing—100 to 200 hours.

Wood construction also requires attention to surface finish, but to a lesser degree than moldless fiberglass. Otherwise, the techniques for building with wood are roughly similar to those used for metal. The parts are the same: ribs, spars, skins, and so on. They are cut and sanded rather than sheared, bent, and hammered, and they are glued together rather than riveted. The number and purpose of the individual steps, though, is similar, though metal has the advantage of requiring no surface finish, if it is carefully executed in the first place.

"Tube and rag," as the most old-fashioned kind of construction is called, requires cutting pieces of chromium-molybdenum steel tubing, carefully shaping their ends, and welding them into a framework which, like a bridge truss, provides the structural strength of the airplane. A lighter framework, usually of wood, is added outside the tubular structure, and serves to give shape to the final covering of fabric. On such airplanes, the wings are usually made of wood, and fabric-covered.

The amount of time and difficulty that building an airplane entails is proportional to the number of parts you have to make. You may think that a lot of simple parts can be made more quickly than a few difficult ones, but the numbers tend to overwhelm every other consideration. This rule

applies not only to comparing different types of construction, but also to different designs within a single medium. For instance, if you look at a partially finished Pazmany PL-2 and compare it with a partially finished Thorp T-18 you will observe that the styles of construction are different; the Pazmany has a finer grain—that is, consisting of a larger number of smaller parts. Engineers have their reasons for choosing different styles. It had been felt that the weight penalty of a simpler, coarser-grained style was prohibitive. More recently, however, designers have shown that putting ease of construction ahead of all other factors does not seriously affect performance, and the very simple, utilitarian, easy-to-manufacture style has taken over both in homebuilts and in the industry, where it is synonymous with lower

production costs.

It is highly advisable to enter into an airplane project with a partner, or with several, and either to cooperate on building one plane, or to build several at once. In either case, the progress is more rapid, and utilization will probably be better in the end. Mass production, even on the smallest scale, keeps costs down, because you need to make tooling only once. For most popular designs, parts like wing ribs, fuselage frames, and so on can be ordered from suppliers who are making the most of their tooling.

Take a close look at the design you're thinking about building before starting. Pictures can be misleading; even more so are the performance claims made by some designs. Few designers of experimental plans take the

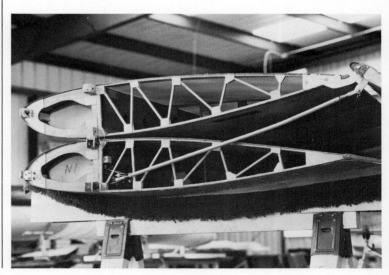

The Mitchell Wing employs styrofoam ribs—sanded to contour after being glued into place—to support a thin-skinned plywood leading edge. Unlikely materials such as styrofoam have proved surprisingly suitable for aircraft structural applications.

trouble to develop meaningful performance tables—a tedious process with often disappointing results. Consequently, the ones they advertise may be a bit "optimistic." The late Ken Rand routinely touted his KR-2 by adding 50 mph to its cruising speed. Exaggeration may take other forms. The Brokaw Bullet, for instance, is said to be suitable for engines of 200 to 500 hp. But the airplane is a huge two-seater with a tiny wing and an empty weight of 1,800 pounds. It doesn't take a genius to guess that with a 200-hp engine it would barely get off the ground. On the other hand, there is no 500-hp aircraft engine on the market; this figure is merely an engineering parameter, meaning that the airframe is structurally capable of handling such an engine.

This kind of mixing of a structural limit with a measured performance point shows up everywhere. An airplane might, for instance, claim a cruising speed of 160 mph and a "maximum speed" of 250 mph. The prospective purchaser imagines that the maximum speed is the fastest he could go in level flight at full power, but it isn't. A plane whose cruising speed is 160 mph will have a top speed of 170 or 175 at the most. "Maximum speed" is, in this case, an airframe limit dive speed—a convention used in design.

Certain rules of thumb are helpful in assessing performance claims. For instance, an airplane's cruising speed is normally 2.5 to 3.5 times

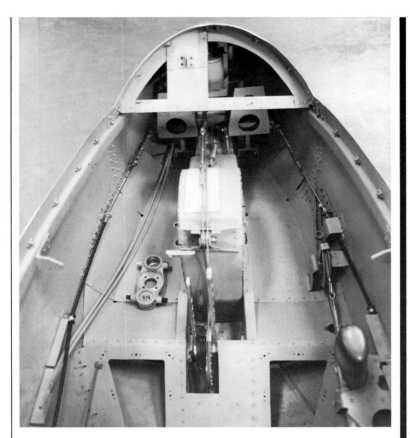

Although it was represented as a 300-hour basement project, the BD-5 is an airplane of great complexity, involving many parts, odd shapes and difficult fits.

its stalling speed. A larger ratio than this is cause for doubt. Especially suspect are very low stalling speeds. Airspeed indicators are unreliable at low speeds, and it is difficult to get good data at stalling speed without calibrated equipment. In more or less good faith a promoter might claim a 40-mph stall, having seen it on his airspeed indicator; but except for the lightest airplanes with the largest wings, attaining such a low stalling speed is improbable.

The maximum cruising speed may be higher in a turbocharged airplane, but then the conditions for realizing those speeds are out of the reach of most amateur fliers. For instance, a turbocharged airplane might attain its maximum cruising speed at 22,000 feet; but that is an altitude at which, for various reasons, one would rarely cruise.

Sometimes performance figures are mathematically deduced from

other figures without being tested. For instance, someone might claim that he can go 200 mph at 22,000 feet on 6 gallons an hour, and so, because he holds 60 gallons, his cruising range is 2,000 miles. In real life, he could never go that far (except with a tailwind), because overall fuel consumption is greater than average en-route consumption (on account of taxi, takeoff, climb, and so on), because he can't carry 10 hours' oxygen in the airplane to breathe at such a high altitude, and because you always need a reserve at landing.

The performance of factory airplanes gives a good sense of what can and what can't be done. Though homebuilts are generally smaller than factory ships, they usually have smaller engines too, and their performance is in the same broad spectrum as that of a Piper or a Cessna. In fact, on at least one occasion the Pazmany Efficiency Contest at Oshkosh— a test of speed, range and fuel efficiency—was won by a big factory Cessna. So you have to be skeptical of claims that a certain airplane outperforms all others by a wide margin. It's possible— but it's far more possible that a man might be stretching the truth.

In the world of homebuilts it is not unusual to advertise performance before the airplane has been flown at all. A couple of years ago a little two-seater appeared at the annual EAA Fly-In at Oshkosh claiming 45-mph stall (which was plausible) and 230-mph cruise (which wasn't) on a turbo-charged VW engine. Quite apart from serious questions about the reliability of such an engine, it can be stated as a virtual certainty that the engine could never propel a two-seat airplane, especially one with a low stall speed and therefore a large wing, at anywhere near that speed. From 140 to 150 would be more likely, perhaps 180 with full blower at high altitude.

Also every year at Oshkosh one sees extremely radical new designs for which extravagant claims are made: roadable planes, ducted-fan standoff scale single-seat jets, and modified V-8 engines or Mazda rotaries, touting remarkable power outputs at low fuel consumptions. All this is pure pie-in-the-sky. Promoters, when they are not unvarnished liars, seem to be able to talk themselves into believing almost anything; and in turn their own faith seems to convince others. It would be most unwise to believe anything you hear from someone who is trying to make money from a project, let alone to give such a person your cash. Only airplanes that have logged 100 hours or more should be taken seriously by a prospective buyer or builder.

The way to get a realistic assessment of the capabilities of a homebuilt design is to see it in flight, talk with *several* builders (not the promoter, and not in his presence) and, if possible, take a ride and try it out for yourself. Although people who are experienced in airplane design and operation can spot an exaggeration or an outright fraud with ease, most laymen can't. They have to be guided by others, so it is essential for them to get the widest selection of the most unbiased opinions. This is a field in which greed and naivete often collaborate to produce remarkable disappointments.

Quite often the maiden voyage is made not on wings, but on the back of a truck.

In spite of a few conspicuous fakes, however, the homebuilt market is certainly not merely a trap for the unwary. From the designs that have been around for years, the process of natural selection has weeded out the unworthy. The more popular types have logged thousands of hours in the air, their characteristics are well known, and their basic designs have been improved to reflect in-service experience, just like the designs of factory products. Impartial information about their construction and their performance is not hard to come by.

First Flight

At the opposite end from selection is test flying. You approach it with heightened feelings, having long ago prepared for the moment that is finally at hand. You

When you are building the airplane, advice and comment are cheap. When you fly it for the first time, you find yourself completely alone.

are full of both misgivings and impatience. Last minute adjustments and finishing touches seem to drag on for weeks. Then one day you trailer your plane to the airport. A small crowd has gathered. You taxi gingerly, fearful at every moment of some humiliating blunder or breakage. The first few times you take the airplane out onto the runway you slowly accelerate to higher and higher speeds, lift off and immediately chop the power; you then glide along for a few hundred feet, letting your plane settle back to the runway, feeling the controls and testing the stability, familiarizing yourself with your plane's behavior.

Then, finally, you decide to actually lift off and climb away for the first bona fide flight, at least around the pattern, perhaps farther afield.

For some pilots this moment is one of incomparable exhilaration. For others it is an anticlimax, empty of the feelings of pride and satisfaction they had expected. Each person reacts differently. But for each the feelings are so many, and blended of such disparate elements—anticipation, fear, impatience and reluctance, satisfaction and disappointment—

that it is hard to precisely decipher the maze of emotions. Some find that the achievement of the dream of years dawns on them only slowly; others feel good right away and go off for a big party. If there is a delay in the full realization of completion, it is because one becomes so attached to the chain of tasks and the pleasant, repetitive activities of building that ending them is slightly unnerving, like graduating from college.

First flights of proven designs are not necessarily hazardous. The danger resides in the pilot, not in the airplane. Pilots of little experience, or ones who have not flown for a long while or who have never flown a very small, light airplane, need time to adjust themselves to the sensations of flying a homebuilt. Pressure from onlookers and friends—unintentional pressure working on the builder's fear of disappointing his audience—can make pilots rush into flight when they would be well advised to postpone it.

The best way to conduct a first flight is to invite someone to come along with you to the airport, a person with a good knowledge of airplanes. Go at a time when a lot of other people aren't around. Proceed very slowly, heed the cautions of your companion, and remember that after spending years building your airplane, you can afford to spend a little time making its acquaintance. Many accidents in homebuilts occur on early flights, when pilots still uneasy in their airplanes—if not downright terrified—lose control of them or are distracted by some unexpected problem, like a canopy popping open on takeoff.

Having established that the new airplane runs reliably, that it responds to the controls and is properly trimmed, you should practice stalls and landings and slow flight, which are potentially the areas of greatest difficulty for the pilot. A series of short familiarization flights, interspersed with periods of discussion and relaxation, are better than exhausting marathon sessions. Slow flying should be done at high altitude, and a parachute should be worn

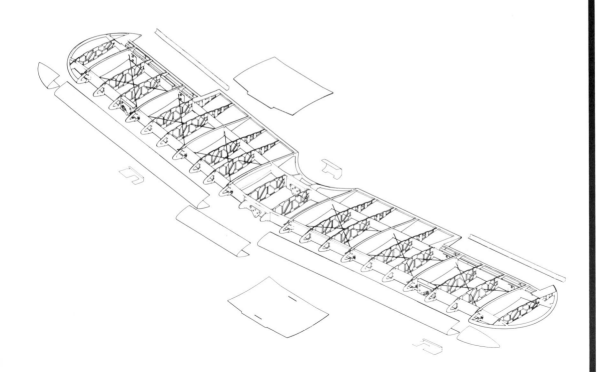

until you have total confidence in yourself and the plane—for months, if need be.

You should be aware of the potential danger of flutter at high speeds, and proceed very cautiously in moving up the speed range. Center of gravity position is also of great importance, and the pre-flight weighing and CG calculation must be done with great care. Follow the advice in the designer's manual to the letter. If there is none, the VariEze first flight manual, available from Rutan Aircraft Factory, is an excellent general guide. Always remember that though airplanes are dangerous, they are not so dangerous as pilots.

The Place To Start:
Although membership in the Experimental Aircraft Association is hardly obligatory, it can be very helpful to someone wanting to build an airplane. Many people join EAA just to participate in

the fantasizing. The annual membership fee, a well-spent $20, includes a subscription to the EAA magazine, *Sport Aviation*, which is an exasperating mixture of excellent technical articles, useful information on problems and solutions, cornball ramblings about flying, inexpert opinion, bad advice, and occasional downright lies. The policy of the magazine, and of the EAA generally, is not to judge the merits of designs or ideas, but to present them all impartially. While a beginner can't hope to separate all the wheat from the chaff, he would do well to keep in mind that the mere fact that a piece of information (or an advertisement) appears in *Sport Aviation* does not mean that

it is correct or honest. The field is, after all, experimental and the EAA has done a good job of keeping it so.

EAA publishes a booklet, "Sport Aircraft You Can Build," which combines some articles chosen from *Sport Aviation* along with answers to frequently asked questions, and a complete catalog of the plans and kits that are available. The choice, as you will find, is astoundingly wide. Another excellent source of information is the Homebuilts supplement to *Jane's All the World's Aircraft;* it can be obtained or ordered at local bookstores, or from Aviation Book Company, 555 W. Glenoaks Blvd., Glendale, CA 91202.

EAA publishes a list of local chapters all over the USA and the world; this gives you a chance to go to a meeting or two, talk with people, see homebuilts in progress or in use, and begin to form some kind of realistic picture of what homebuilding involves. Most chapters meet once a month or more, and divide their time among progress reports by builders, instructive talks, lectures from local FAA representatives, and the ravings of harmless madmen.

Besides chapter meetings, there are local and regional Fly-Ins, some of them very large, which take place at least annually. At the big ones designers and sellers come from all over the country to promote their wares. Also promoting their wares are various ironmongers who capitalize upon the excitement generated by the event to sell, say, two cracked right gear struts to somebody who thinks that by buying some parts he is beginning a project.

And then every summer there is "Oshkosh," held on Wittman Field in Oshkosh, Wisconsin. By most accounts this is the best airshow in the world. Although it lacks the participation of the military and transport aircraft manufacturers who take their wares to the Paris, Hanover and Farnborough shows, the annual EAA Fly-In makes up for their absence with an otherwise complete array of aircraft, from antiques to jets. Hundreds of homebuilts appear. Partially completed airplanes are trailered in by their proud builders. Each day swarms of homebuilts compete for the crowd's attention in the fly-by pattern. Thousands of factory-built aircraft fly in; hundreds of thousands of visitors attend the week-long show, swamping the local campgrounds and highways. Oshkosh gets its pick of airshow performers, too, with breathtaking displays of aerobatic skills. Seminars and demonstrations for builders go on continuously, several at a time in different tents, and commercial manufacturers of parts, equipment and airplanes set up elaborate exhibits. Oshkosh is a unique combination of trade show, convention, vacation resort and gathering of the clans, an unforgettable encounter with the beautiful products of human ingenuity, craftsmanship, patience and ambition.

To join the EAA, send the $20 membership fee to: EAA, Post Office Box 229, Hales Corners, Wisconsin 53130. If you are interested in building an airplane, this is the place to start.

Aileron
A control surface, normally located on the outboard edge of each wing, used to bank the airplane for turns.

Airframe
The entire structure of an airplane.

Aspect Ratio
The ratio of the length of a wing, tip to tip, to its breadth, front edge to back edge. A high aspect ratio is associated with efficient flight at low speeds; but as speeds increase, lower and lower aspect ratios are used because they permit stronger, stiffer wings.

Center of Gravity
The balance point of the entire airplane, normally placed about a quarter of the way back from the front edge of the wing. In a canard airplane, however, the center of gravity is somewhere between the front and rear wings.

Chord
The distance from the front edge to the rear edge of the wing. A rectangular wing has a single chord, but a tapered wing has a root chord, a tip chord, and a "mean aerodynamic chord," a fictional number used in aerodynamic calculations.

FAA
Federal Aviation Administration. The branch of the federal government charged with the protection and regulation of aviation.

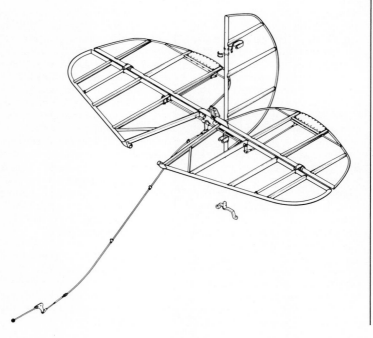

Flutter
Vibration of large airframe parts, caused by structural resonance, and often responsible for sudden in-flight breakup.

Gross Weight
The maximum loaded weight of an airplane.

IFR
Instrument Flight Rules. Colloquially used to refer to blind flying in bad weather with the assistance of radio navigation and of ground-based radar controllers.

Leading Edge Radius
The front edge of a wing is rounded. The smallest radius of curvature used is called the leading edge radius, and it gives some indication of the stalling characteristics of the airplane. Generally, the blunter the wing—the larger the leading edge radius—the more gentle and indistinct will be the stall.

Lift Distribution
The spanwise distribution of the upward force produced by a wing moving through the air. Usually lift is greatest near the middle of the wing, and tapers off near the tips. Lift distribution, like leading edge radius, gives designers clues about the stalling characteristics of a wing.

Payload
The weight an airplane can carry. Usage varies, but payload usually *excludes* fuel, because fuel doesn't "pay" in the sense that cargo or passengers do.

Rogallo
A triangular nonrigid wing consisting of a frame of aluminum tubing supporting a sail-like fabric lifting surface.

Span
The length of a wing from tip to tip. Also called "wingspan."

Span Loading
The airplane weight divided by the wingspan. A parameter of the power required for climb; higher loading, more power required.

Stabilator
A type of horizontal tail surface which consists of a single all moving slab, rather than a moveable surface hinged to a fixed surface.

Stall
Below a certain minimum speed—which varies from one airplane to another—the wing abruptly produces much less lift, and the airplane won't stay airborne unless its speed is increased. This sudden loss of lift is called the "stall." It has nothing to do with the engine, as in a stalled car.

Turbocharger
At high altitudes the air is less dense than at sea level, and less oxygen is therefore available for the engine to burn with fuel. Hence performance deteriorates with increasing altitude. A turbocharger is an air pump driven by exhaust gases which forces compressed air into the engine, maintaining sea-level power up to high altitudes.

Useful Load
Normally, payload plus fuel. Or, simply, gross weight minus empty weight.

VFR
Visual Flight Rules. Used to describe flying conditions in which the weather is good, and flight plans or ground control are not necessary.

Wing Loading
The weight of an airplane divided by the wing area. A parameter of landing speed; higher wing loadings mean higher landing speeds.

Wing Twist
To control stalling characteristics, wings are sometimes built with twist, so that the tips are at a slightly lower angle to the oncoming air than the roots. Also often called "washout."

VariEze
Rutan Aircraft Factory
Post Office Box 656
Mojave, CA 93501

Bensen Gyrocopter
Bensen Aircraft Co.
Post Office Box 31047
Raleigh, NC 27612

BD-5
Bede Micro Aviation
456-A Willow Street
San Jose, CA 95110

Dyke Delta
John Dyke
2840 Old Yellow Springs Road
Fairborn, OH 45324

Sonerai
Monnett Experimental Aircraft
955 Grace St.
Elgin, IL 60120

Mitchell Wing
M Company
1900 South Newcomb
Porterville, CA 93257

Osprey II
Osprey Aircraft
3741 El Ricon Way
Sacramento, CA 95825

Christen Eagle II
Christen Industries, Inc.
1048 Santa Ana Valley Road
Hollister, CA 95023

Scorpion II
Rotorway Aircraft, Inc.
14805 S. Interstate 10
Tempe, AZ 85284

Quickie
Quickie Enterprises
PO Box 786
Mojave, CA 93501

Thorp T-18
Thorp Engineering Co.
PO Drawer T
Lockeford, CA 95237

E.A.A.
National Headquarters
Post Office Box 229
Hales Corners, WI 53130